Hypobaric Storage in Food Industry

Hypobaric Storage in Food Industry

Advances in Application and Theory

Stanley P. Burg
Consultant, Miami, FL, USA

AMSTERDAM • BOSTON • HEIDELBERG • LONDON
NEW YORK • OXFORD • PARIS • SAN DIEGO
SAN FRANCISCO • SINGAPORE • SYDNEY • TOKYO

Academic Press is an imprint of Elsevier

Academic Press is an imprint of Elsevier
32 Jamestown Road, London NW1 7BY, UK
The Boulevard, Langford Lane, Kidlington, Oxford, OX5 1GB, UK
Radarweg 29, PO Box 211, 1000 AE Amsterdam, The Netherlands
225 Wyman Street, Waltham, MA 02451, USA
525 B Street, Suite 1900, San Diego, CA 92101-4495, USA

First published 2014

British Library Cataloguing-in-Publication Data
A catalogue record for this book is available from the British Library

Library of Congress Cataloging-in-Publication Data
A catalog record for this book is available from the Library of Congress

ISBN: 978-0-12-419962-0

For information on all Academic Press publications
visit our website at **store.elsevier.com**

Working together
to grow libraries in
developing countries

ELSEVIER   Book Aid
            International

www.elsevier.com • www.bookaid.org

CONTENTS

ACKNOWLEDGMENTS

Hypobaric Storage in Food Industry: Advances in Application and Theory is dedicated to my wife, Monika, without whose motivation and patience the project could never have been completed. I also wish to give special appreciation to my friends and associates, Dr. Duane R. Schultz, Prof. David R. Dilley, and Prof. Thomas L. Davenport, who volunteered to proofread this manuscript and offered suggestions for improvements. Also, to Dr. Davenport for his assistance in hypobaric research, to Dr. Xianzhang Zheng for his enthusiastic support of hypobaric research in China and coauthoring two publications, and to Jed and Dick Bothel (Atlas Technologies) for fabricating Vivafresh warehouses in the United States and allowing me to remotely monitor their performance.

Units: 1 atmosphere = 101.3 kPa = 1000 mbar = 760 mmHgA (absolute) = 760 torr. To simplify comparisons between atmospheric and subatmospheric pressure, gas and vapor concentrations are expressed as percent [gas or vapor] where 2% [gas or vapor] refers to a partial pressure of 0.02 atm.

Abbreviations: low-pressure storage (LP); controlled atmosphere storage (CA); modified atmosphere storage (MA); normal atmosphere storage (NA); volatile organic compound (VOC); gas chromatography (GC); ultraviolet (UV); internal ethylene concentration (IEC); internal carbon dioxide concentration (ICC); respiratory quotient (RQ); respiratory inversion point (IP); mmHg absolute (mmHgA); cloud condensation nuclei (CCNs); 1-methylcyclopropene (1-MCP); relative humidity (RH); basic-Helix-Loop-Helix-PAD (bHLH); ethylene polyethylene (EPE); fibroblast growth factor (FGF); fibroblast growth factor receptor (FGFR); Food and Drug Administration (FDA); hypoxia-inducible factor (HIF); International Commission on Microbiological Specifications for Foods (ICMSF); low-density polyethylene (LDPE); Metabolic System for Disinfestation and Disinfection (MSDD); National Plant Protection Organization (NPPO); Senescence-associated genes (SAGs); ethylene forming enzyme (EFE); soluble cuticular lipid (SCL); headspace gas chromatography (HSGC); discontinuous gas exchange (DGS); passive suction ventilation (PSV); corona discharge (CD).

Knowledge is not knowledge until someone else knows that one knows (Lucillius: Fragment, circa 125 BC)

CHAPTER 1

History of Hypobaric Storage

- 1966: First description of the hypobaric storage method (Burg and Burg).
- 1967: Low-pressure storage patent by Burg issues and is assigned to the United Fruit Co.
- 1968–1971: Fruehauf Corp. builds two prototype LP intermodal containers for United Fruit Co.
- 1972: Dormavac Corp. purchases LP patents and prototype containers from United Fruit Co.
- 1974: Dormavac's assets sold to Grumman Allied Industries.
- 1976: Patents by Burg extend the LP storage pressure range to 4.6–20 mmHg.
- 1979: Grumman Corp. and Armour & Co. awarded the US Food Technology Industrial Achievement award for developing hypobaric transportation and storage systems.
- 1984: Premature introduction of Grumman's Dormavac container into commercial service results in mechanical failures and spoilage delivering hanging-lamb from Australia to Iran.
- 1985: Grumman sells its hypobaric technology. Academic publications claim that hypobaric storage is a flawed technology.
- 1987: Patents by Burg issue describing a "dry" hypobaric process that stores respiring plant matter (US 4,685,305) and nonrespiring animal matter (US 4,655,048) without mechanical humidification.
- 1989: Two prototype VacuFresh hypobaric intermodal containers embodying the "dry" hypobaric concept are fabricated by Welfit Oddy (Pty) Ltd in Port Elizabeth, S. Africa, and ABS, Lloyd's, and ISO tested and certified.
- 1991: IRS files a complaint for underpayment of taxes vs. the patent licensee under US 4,655,048 and 4,685,305. Commercial development of LP ceases.
- 1997: Burg abandons US 4,655,048 and 4,685,305 to cancel the patent licenses.

Hypobaric Storage in Food Industry. DOI: http://dx.doi.org/10.1016/B978-0-12-419962-0.00001-2

- 1999: Welfit Oddy (Pty) Ltd is sold and the new owner does not want to attempt to market the VacuFresh container. Burg retains ownership of the VacuFresh blueprints.
- 2007–2009: Publications by Burg and Zheng demonstrate that Western academic literature claiming hypobaric storage is a flawed technology was based on experimental errors. Chinese publications subsequently describe favorable LP results with more than 60 different commodities.
- 2009: A. Colombian flower grower purchases a hypobaric warehouse from Atlas Technologies, Port Townsend, WA, after learning that for 10 years Burg's former patent licensee had clandestinely stored flowers for holiday price appreciation in Dormavac hypobaric intermodal containers refurbished to operate in accord with US Patent No. 4,685,305.
- 2010: Provisional patent application serial No. 61/170,506 entitled "Systems and Methods for Controlled Pervaporation in Horticultural Cellular Tissue" is filed by Burg et al. and assigned to Atlas Technologies.
- 2010: The Science and Technology Committee of Shanghai issued a grant to develop a hypobaric warehouse.
- 2011: The Science and Technology Committee of the People's Republic of China provided funding to build a hypobaric storage unit for use onboard warships. First Chinese hypobaric warehouse is sold and put into operation.
- 2012. Colombian flower grower orders a second Vivafresh hypobaric warehouse.
- 2012: Burg files provisional patent application entitled "Controlled and Correlated Method and Apparatus to Limit Water Loss from Fresh Plant Matter During Hypobaric Storage and Transport" (US Provisional Patent Application No. 61705016). US patent issues in 2014.

Three decades ago laboratory studies demonstrated remarkable increases in the storage life of horticultural crops, meat, fish, poultry, and shrimp kept in a water-saturated flowing low-pressure (LP) atmosphere (Burg, 1967, 1973, 1975,1976; Burg and Burg, 1966; Tolle, 1969, 1972; Dilley, 1977, 1978; Dilley and Dewey, 1973; Dilley et al., 1975; Apelbaum et al., 1977a,b,c; Salunkhe and Wu, 1973, 1975; Spalding and Reeder, 1976a,b, 1977; Spalding et al., 1978; Alvarez, 1979, 1980; Mermelstein, 1979; Jamieson, 1980a,b; Bangerth, 1973, 1974, 1984; Table 1.1). Between

Table 1.1 Maximum Storage Life (days) in NA, CA, and LP			
Commodity	NA	CA	LP
Asparagus	14–21	21 + (slight benefit)	42
Avocado (Lula)	30	42–60	102
Banana	14–21	42–56	150
Carnation flower	21–42	No benefit	140
Cucumber	9–14	14 + (slight benefit)	49
Green bean	7–10	14	38
Green pepper	14–21	No benefit	50
Lime (Persian)	14–21	Juice loss, peel thickens	90+
Mango (Haden)	14–21	21 + (slight benefit)	56
Mushroom	5	6	21
Papaya (Solo)	12	14	28
Pear (Bartlett)	60	100	200
Protea (flower)	7	No benefit	30+
Rose (flower)	7–14	No benefit	60
Spinach	10–14	14 + (slight benefit)	50
Strawberry	7	7 + (off-flavor)	21

Copyright @ 2004 by CABI. Table (slightly updated) from the Preface of Burg (2004) has been reproduced by STM permission from CABI to Elsevier (Burg, 2004, ISBN0 8 5199 011).

1976 and 1982, prototype Grumman/Dormavac hypobaric intermodal containers (Figure 1.1) successfully shipped horsemeat from Texas to France, Belgium and Italy, pork from South Dakota to Hawaii, asparagus from the Dominican Republic to New York City, mangos from Mexico to Japan, and hanging-lambs from Australia to Miami (Burg, 2004). Solo papayas exported from Hawaii to Los Angeles and New York City in Dormavac containers remained firm and nearly disease-free, and after transfer to atmospheric pressure ripened with excellent flavor, texture, aroma and shelf life. In companion shipments, papayas from the same source, transported to the same destinations in conventional refrigerated intermodal containers, softened prematurely in-transit and developed a high incidence of peduncle infections, stem end rot, and surface lesions[1] (Alvarez, 1979, 1980; Table 1.1). After 55 days at sea, hanging-lambs shipped in a Dormavac container from Australia to Iran were rated by an Australian Veterinary attaché to be equivalent in appearance to 2- to 3-day old chilled carcasses (Husband, 1982; Sharp, 1985). An Armour Co.

[1]Postharvest papaya losses of up to 75% due to desiccation and poor ripening are often reported to Hawaii's shippers by mainland US wholesalers and retailers.

Figure 1.1 Grumman/Dormavac 12.2 m (40 ft) intermodal hypobaric container. Single door is drawn in by four cam operators and suspended with slotted hinges to assist alignment. Copyright @ 2004 by CABI. Permission granted via SMT from CABI to Elsevier to reproduce Figure 13.4, Grumman/Dormavac 10.2 m (40 ft) intermodal hypobaric container (Burg, 2004, ISBN0 8 5199 011).

meat-tasting panel found such lamb to be unusually tender, with the texture and intense flavor typical of "dry-aged" meat, without the sour taste, sogginess, and other abnormalities associated with vacuum-packaged "wet-aged" meat (Burg, 2004).

Dormavac containers consumed excessive power and were too expensive and heavy for general use (*Chapter 13*), but would have been profitable transporting hanging-lambs to the Middle East where meat in that form is required for religious reasons. Parvese Ltd. purchased 100 containers for that purpose, but en route from Australia to Iran the first commercial shipment encountered severe weather and heavy seas. A Grumman engineer described the result:

A critical mistake was made in Grumman's Dormavac Program when the first commercial containers were shipped out prematurely by Management/ Marketing against strong objections made by the Engineering Department.

The rest is history: we had major failures in our equipment components and major leaks in our glycol and freon plumbing, causing a loss of approximately 80% of the lamb being transported in five containers from Australia to Iran. This premature introduction of the container into service and the subsequent failures led to cancelation of 100 units that were sold to Parvese Ltd. This further led to a total demise of the Grumman Dormavac program which was never able to overcome the disapprobation.

Even before Grumman terminated its Dormavac Program, and during subsequent years, peer-reviewed academic publications described inherent problems with the hypobaric storage method (Tolle, 1972; Wu et al., 1972a,b; Sharples, 1974; Kader, 1975; Stenvers and Bruinsma, 1975; Salunkhe and Wu, 1973, 1975; Lougheed et al., 1977, 1978; Brisker, 1980; Hughes et al., 1981; Theologis et al., 1992; Abeles et al., 1992; Goldschmidt et al., 1993; Paul et al., 2004; Lagunas-Solar et al., 2006; An et al., 2009; Thompson, 2010). The following is a sampling of criticism paraphrased from these sources:

Hypobaric equipment is complex, unreliable, and wasteful of energy. Low-pressure technology has implicit technological deficiencies, insoluble economic problems, and presents an insurance and implosion problem. Water loss due to rapid diffusive escape of moisture at a low pressure causes severe commodity desiccation. Hypobaric storage is unable to prevent senescence and other C_2H_4-induced responses by lowering the intercellular C_2H_4 concentration because "active" C_2H_4 is bound to its receptor. LP is simply CA storage lacking the benefit of added CO_2. Plant commodities are injured by rapid pressure changes. A low pressure induces "hypoxic stress," causing fruits to produce wound C_2H_4. In a water-saturated low-pressure environment, horticultural commodities express genetic drought–stress responses associated with the perception of desiccation. Few crops can tolerate the level of hypoxia required to prevent decay induced by the high humidity required for hypobaric storage. A low pressure results in poor quality and unsatisfactory ripening because it evacuates and outgases flavor and aroma components from fruits.

The academic belief that hypobaric storage is a flawed technology originated from experimental errors in LP research caused by non-precise temperature control, cold-spots on the vacuum chamber's surface, humidifying at atmospheric pressure rather than a low pressure, inadequate air changes, leaky vacuum chambers, and a failure to realize that the high turgor pressure of plant cells prevents LP from causing volatiles to boil and "outgas."

Postharvest researchers and peer reviewers were unaware of these mistakes because they had only a meager understanding of vacuum technology and heat and mass transfer at a low pressure. When the Dormavac program was terminated many academicians conjectured that if Grumman could not make a commercial success of LP after investing so much time, effort, highly skilled engineering talent, and nearly $100,000,000, this confirmed their belief that the hypobaric process was "too complicated to ever work" and had inherent defects that could not be remedied. This critical literature so-diminished interest in hypobaric storage that during the past 20 years only a few LP studies have been published in the West.

US and International Patents issued in 1987 describing a "dry" hypobaric process that stored respiring plant matter (US 4,685,305) and nonrespiring animal matter (US 4,655,048) without mechanical humidification. Expanded dry air (<1% to 2% RH) continuously enters the vacuum chamber through a pneumatic air horn (Figure 6.1), and the air-change rate is adjusted so that moisture generated by the minimum rate of transpiration needed to remove respiratory heat from the stored plant matter continuously saturates the storage atmosphere. Prototype VacuFresh containers embodying the "dry" concept, rated for full internal vacuum and applicable handling/transportation load factors in accordance with ANSI MH5.1.2 M and ISO 1496-3, were fabricated in South Africa by Welfit Oddy (Pty) Ltd (Figure 1.2). These units met the racking-test criteria for ABS and Lloyd's certification and were cost effective, as energy efficient and nearly as lightweight as NA or CA intermodal containers, more cost effective than air transport, and their cost advantage vs. air transport has increased since the price of fuel has continuously risen and air transport is less fuel efficient than sea transport.

In 1991, the IRS filed a complaint for underpayment of taxes vs. Gelco Internacional SA, the sole patent licensee under US 4,655,048 and 4,685,305, and would not permit Gelco's ongoing expenditures for the hypobaric project to take precedence over past due taxes. Commercial development of hypobaric storage ceased until 1997, when Burg abandoned the "dry" LP patents to cancel Gelco's patent license without litigation. Welfit Oddy (Pty) Ltd was sold and the Company's new owner did not wish to risk funds attempting to develop and market an unproven hypobaric container.

In 2009, a Colombian flower grower ordered the first "Vivafresh" hypobaric warehouse (patent pending—Burg et al., 2009) from Atlas Technology, Port Townsend, WA, USA, after learning that for nearly

Figure 1.2 Upper: Prototype VacuFresh intermodal container photographed during fabrication at Welfit Oddy (Pty) Ltd, in Port Elizabeth, South Africa. The author is inspecting the interconnected stiffening rings and reinforcing ring gussets. A huck-bolted mounting clip box is welded at the center of the upper end frame. Longitudinal channels protect the tank from physical damage and are not structurally required. Lower left: Equipment end after installation of insulation and cladding. Copyright @ CABI (2004). Figure 11.12 in Burg (2004, ISBN 0 85199 8011 1) (Lower right). Photograph of door end, courtesy of Mike Spearpoint. T. ENG (CEI).

a decade Gelco Internacional SA had covertly stored flowers for holiday price appreciation in Dormavac containers refurbished to operate in accord with US 4,685,305. The Colombian planned to airship roses to Miami at times when there was so little demand for them that their market price was very low and air-carriers discounted the cost of transportation. He hoped to store the flowers in Miami until their retail price increased up to 10-fold on a holiday or other occasion, at which time the cost of air-transporting flowers to Miami peaked. The Vivafresh warehouse (Figure 1.3) has reliably preserved the initial vase-life of roses during the longest storage attempted, 7 weeks, confirming research published nearly 40 years ago (Dilley et al., 1975; Table 1.1). The roses have been sold at a premium price in Miami supermarkets, while roses imported from Colombia in the same air shipments lost storage and vase-life so rapidly in the grower's state-of-the-art high-humidity refrigerated room that they had to be discarded within 5 days. After 7 weeks in the Vivafresh warehouse, *Alstroemeria* flowers were transported by refrigerated truck from Miami to Boston and sold without claims. The Vivafresh warehouse's cost was recovered and a profit generated within 2 years. In 2012, the Colombian grower purchased a second warehouse. This success story has renewed interest in hypobaric warehouses, and in exporting mangos, papayas, asparagus, avocados, flowers, and other crops in VacuFresh hypobaric intermodal containers.

Mangos are one of the world's most abundant fruit crops, and nearly 10% of the world's production is exported by air or in

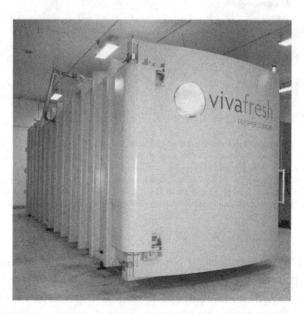

Figure 1.3 Vivafresh 3332 ff³ (94 m³) hypobaric warehouse. Internal dimensions = 8.3 × 10 × 40 ft³ (2.5 × 3 × 12.2 m³). Photograph courtesy of Dr. Tom Davenport, 2013.

intermodal sea containers. To minimize softening and spoilage during a 2−3 week ocean voyage, ripening must be suppressed by harvesting mangos at a half-mature stage[2] or earlier. This requirement reduces the grower's production by 20−50% of the weight gain various mango varieties would have experienced if they had been allowed to reach full maturity during 1−2 additional months in the field. Because less-mature mangos ripen slowly with poor texture, flavor, color, and aroma, the grower forfeits the premium price he might have received for a higher quality product and the consumer never experiences the exotic flavor and aroma of a fully mature ripe mango. The balance between harvest maturity, reliable sea transport, total field production (weight), and the fruit's final quality and selling price determines the grower's profit from his crop.

Inhibiting ethylene action with 1-methylcyclopropene (1-MCP) only extends mango storage life by 1−2 days (Slaughter, 2009). Depending on their maturity, mangos can be preserved for 14−21 days in NA (Hardenburg et al., 1986; Welby and McGregor, 1997). Storage life may be extended for up to 6 additional days by irradiating fruit at 25 krads for insect control (Dharkar et al., 1966; Section 11.9). Half of "immature" mangos shipped in Purfresh ozone-generating intermodal containers ripen and soften in-transit within 22 days (immpurReportNFLFoodSafety0410.pdf). Intermodal CA containers equipped with ethylene scrubbers can only preserve half-mature mangos for 2 weeks, after which the fruit has little shelf-life (Bender et al., 2000; Ullah et al., 2010a,b). Fully mature mangos remain firm during 8 weeks in LP. They can be reliably exported by sea in a VacuFresh container to any destination in the world, and after transfer to atmospheric pressure ripen with excellent shelf-life, flavor, texture, color, and aroma (Davenport et al., 2006) (Figure 1.4).

Western scientists may be motivated to reconsider hypobaric storage because of recent Chinese publications describing favorable LP results with bayberry; blueberry; waxberry; mulberry; blackberry; gooseberry; persimmon; honey, juicy, Dajiubao and Datuanmilu peaches; loquat; "Quanxing" apricots; tomatoes; bananas; Dongzao, Zizyphus, and Lizao Jujube; Angelo and Whangkeumbae plums; Gold

[2]Mango maturity is judged by the development of the fruit's shoulders relative to its stem. Fully mature = outgrown shoulders; half-mature = shoulders in-line with the stem; immature- = shoulders below the stem.

Figure 1.4 Hypobaric warehouse (50 m³) designed by Dr. Xiangzhang Zheng and the Shanghai Kind-Water Preservation Technology Company (Chinese patent application No. 11554941, Zheng and Zheng, 2009c). Photograph courtesy of Dr. Zheng.

and Ciguan pears; Huanghua peas; strawberries; cherries; lychees; dates; Jintong cucumbers; asparagus; Cangshan garlic scape; mushrooms; *Tricholoma matsutake sing*; Aceroe shallots; hollow vegetables; lettuce; lotus root; bamboo shoots; green snap beans; blue onions; celery; brocolli; yams; potato; apples; water radish; spinach; Quing vegetables; Jimao vegetables; swamp cabbage; China Crucian fish; Luo shrimp; beef; pork; chicken; liver; Coba; and cooked foods such as steamed breads, flapjacks, bean curd, Kaofu and Boye (Chang, 2002; Chang and Wang, 2003; Chang et al., 2004; Wang, 1991, 2006; Wang and Zhang, 2008; Wang et al., 2001, 2004a,b,c, 2007, 2008; Yao et al., 2009; Li and Wang, 2007; Li et al., 2004, 2005a,b, 2006, 2007a,b, 2008, 2009, 2010; Li and Zhang, 2006; Li and Zhou, 1993; Sun and Li, 2009; Huang and Zhang, 2001, Huang et al., 2003; Han and Zhang, 2006; Han et al., 2006; Hao and Wang, 2004; Zhang et al., 2005a,b; Zheng and Xiong, 2009a,b; Jiang et al., 2009; Chao et al., 2004; Chen et al., 2004a,b, 2005a,b, 2006, 2007, 2008, 2013; Xue et al., 2003a,b; Gao et al., 2006, 2008; Tao et al., 2003; Cao et al., 2004, 2005; Zhou and Liu, 2004; Zhou et al., 2007; Yang et al.,

1993, 2010; Kang and Zhang, 2001; Wang et al., 2001, 2004a,b, 2007, 2008; Mingli, 2001; Yao et al., 2009; Qu et al., 2005; Burg and Zheng, 2007, 2009; Zhou and Liu, 2004; Zhou et al., 2010; Zou et al., 2011; Zheng and Zheng, 2008, 2009a,b; Zheng et al., 2011;Wu et al., 2011; Fu, 2010; Hu et al., 2012; Wenxiang et al., 2006; Yi, 2010).

The Science and Technology Committee of Shanghai issued a grant in 2010 to assist in developing a hypobaric warehouse. In 2011, the Science and Technology Committee of the People's Republic of China provided funds to build a hypobaric storage unit for use onboard warships. Research units (www.cr-expo.com) are available in China, a system for a warship has been completed, and in 2011 a 50 m^3 hypobaric warehouse was sold in China to store fish. Hypobaric storage research is being carried out at the Institute of Naval Medical Science in China (Section 12.6). In 2009, Zheng and Zheng filed Chinese patent 2009100510253 on an LP storage device. In 2012, the Quartermaster Institute's Department of General Logistics filed five joint patents on LP storage vehicles with the Shanghai Kind-Water Preservation Technology Company, and in 2013 they filed a patent on the hypobaric cold chain for preserving fresh-cut vegetables.

Hypobaric storage is not simply a form of CA storage lacking the benefit of added [CO_2], as many Western postharvest physiologists have claimed and some continue to believe (Lougheed et al., 1977, 1978; Anon, 1974; Kader, 1975; Abeles et al., 1992; Spalding and Reeder, 1976b; Tolle, 1969; Stenvers and Bruinsma, 1975; Goldschmidt et al., 1993; Frenkel and Jen, 1989; Larue and Johnson, 1989). The hypobaric process prevents all major causes of postharvest losses during storage, transport, and distribution, including commodity desiccation (Sections 2.1–2.3, 4.3; Chapter 10); ethylene production, accumulation, and action (Section 3.5); bacterial and fungal decay (Section 12.5); low-[O_2] injury (Chapter 8); high-[CO_2] injury (Section 3.2); physiological disorders (Wang and Dilley, 2000; Section 3.1); and LP may kill surface-feeding insects (Chapter 11). CA provides little or no benefit preserving commodities such as roses, mangos, papayas, and asparagus, and yet LP extends their storage life manyfold (Reid and Jiang, 2005; Bender et al., 2000; Burg, 2004; Table 1.1).

Chapter 2 describes experimental errors in LP academic publications. Chapters 3 and 4 explain why horticultural commodity storage is improved when gas and vapor mass transport and heat transfer take place at a low pressure. This background information and a review in Chapter 5 of "materials and methods" used in laboratory LP research will help the reader to comprehend hypobaric effects that eluded detection for more than 45 years and are revealed for the first time in Chapters 6–10.

CHAPTER 2

Experimental Errors in Hypobaric Storage Research

Problems with hypobaric storage and Grumman's Dormavac intermodal container were summarized in an article entitled "LPS—Great expectations" (Lougheed et al., 1977). Initial cost and royalties, insurance against implosions, desiccation caused by chamber leakage or a failure of the humidification system, power interruption, breakdown of the refrigeration system, inadequate heat transfer, excessive energy consumption, a high cost, produce injury due to rapid pressure changes or a low enough [O_2] concentration to prevent microbial growth and sporulation, an inability to control pathogens due to the high humidity that is required for LP storage, and new disorders were identified as defects in hypobaric storage or areas of concern. Burg (2004) and Burg and Zheng (2007, 2009) reviewed and evaluated these and other criticisms of LP storage.

Academic skepticism regarding low-pressure storage originated from experimental errors in Western studies published prior to 1985. This literature so-diminished interest in LP that hypobaric research essentially ceased at that time. None of the erroneous reports have been retracted, and the studies still are frequently cited. Consequently, many contemporary plant physiologists are unaware that hypobaric storage provides unusual benefits without the claimed disadvantages. In his book on CA storage, Thompson (2010) refers to a flawed study carried out 31 years ago by Hughes et al. (1981) as evidence that "a major engineering problem with hypobaric storage is that the lower the atmospheric pressure the lower the boiling point of water, which means that the water in fruits and vegetables will be increasingly likely to be vaporized." Water cannot boil during hypobaric storage (Table 9.1), and the amount of water vaporized from plant matter depends on the availability of latent heat, not on the pressure (Section 4.3). A commodity stored in a refrigerated space cannot remain at a constant temperature and lose more water than its

Hypobaric Storage in Food Industry. DOI: http://dx.doi.org/10.1016/B978-0-12-419962-0.00002-4

respiratory heat is capable of vaporizing unless it is receiving heat from the environment (Gac, 1956). Likewise, in a Produce News interview (Niblosk, 2012), Staby and Reid refer to LP as "likely to cause an excessive water loss." Staby last carried out LP research nearly 30 years ago (Staby et al., 1984). These researchers are unaware that during the past three decades, all factors capable of increasing water loss from plant matter during LP storage have been identified, and methods devised that reduce commodity water loss to a theoretical minimum (Sections 2.1, 2.3, 9.3).

More than 30% of fresh horticultural commodities spoil during transport and distribution (Kader, 2005; Saraswathy et al., 2010; U.N. Food and Agriculture Organization), and aside from mechanical injury and improper harvest time, water loss is usually the most important cause. The amount of water that can be lost before produce becomes unsalable varies from approximately 3% in lettuce to 10% in cabbage, celery, and cut flowers (Robinson et al., 1975; Burton, 1982; Agrotechnology and Food Science Group, Wageningen), and in most commodities is 5–7%. Many publications have reported that hypobaric storage causes an excessive water loss, while in other studies water loss from the same commodity has been negligible at an ostensibly identical hypobaric condition. How can this be explained?

2.1 A LEAK IN AN LP CHAMBER INCREASES COMMODITY WATER LOSS

Leakage of ambient air into the LP vacuum chamber causes an excessive commodity water loss during hypobaric storage. Researchers and peer reviewers have underestimated the difficulty of constructing a leak-tight laboratory vacuum apparatus, and the effect that leakage has on commodity water loss during LP storage. The lower the pressure the more in-leaking air expands, increases in volume and decreases in humidity as it enters an evacuated storage chamber. At a lower pressure leakage provides a greater part of the air change flowing through the system. There is no obvious indication that leakage is occurring because the LP system's pressure controller automatically compensates for leakage and maintains the selected vacuum, and the commodity's transpiration keeps the storage humidity near saturation. Although several publications cautioned that leakage into a hypobaric apparatus causes commodity water loss (Lougheed et al., 1977, 1978;

Burg, 2004, 2005; Burg and Kosson, 1982, 1983), the chamber leak rate was never measured and reported in a published Western laboratory hypobaric storage study prior to 2005.

The leak rate is determined by evacuating the system to a low pressure, isolating it, and measuring the mmHg/h rate at which the pressure rises. Until the low-pressure to atmospheric-pressure ratio exceeds 0.53, the leakage flow is "choked" (turbulent), the air mass leaking-in each hour depends only on the upstream atmospheric pressure, not on the downstream storage pressure, and the rate of pressure rise is constant. A hypobaric plant growth chamber designed by NASA to simulate a Martian greenhouse had a 27.8 mmHg/h (3.7 kPa/h) leak rate (Corey, 2000), and systems used for laboratory LP studies are likely to leak as rapidly or to an even greater extent unless they have been specially fabricated to prevent leakage. If NASA's chamber was operated without commodity present under the conditions indicated in Figure 2.1, flushed with one saturated air change per hour, the relative

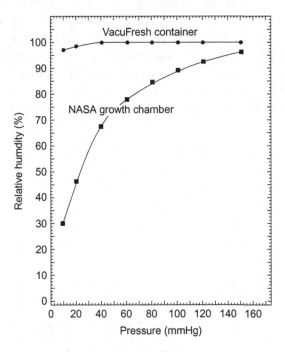

Figure 2.1 Relative humidity computed at various operational pressures inside NASA's experimental hypobaric growth chamber (Corey, 2000) and in a VacuFresh hypobaric intermodal container (Burg, 2004, 2005). Operating conditions: leak rate = 3.7 kPa/h (27.8 mmHg/h) in the NASA chamber, 0.04 kPa/h (0.3 mmHg/h) in the VacuFresh container; ambient condition = 25° C, 50% RH; flowing one water-saturated, 0°C, low-pressure air change per hour; chamber temperature = 0° C.

humidity would only be 30% at 1.33 kPa (10 mmHg) and 85% at 10.67 kPa (80 mmHg). This same protocol in an empty VacuFresh hypobaric intermodal container (leak rate = 0.04 kPa/h = 0.3 mmHg/h) creates a 97% RH at 1.33 kPa, and 100% RH at or above 5.33 kPa (40 mmHg). "Medium" vacuum chambers with 0.004−0.05 kPa/h (0.03−0.4 mmHg/h) leak rates sometimes have been constructed for NASA studies (Corey et al., 2002; He et al., 2003). Grumman/ Dormavac hypobaric intermodal containers had leak rates as high as 1.33 kPa/h (10 mmHg/h). The leak rate in an Atlas Technology Vivafresh hypobaric warehouse (Figure 1.3) is 0.093 kPa/h (0.7 mmHg/h), and the rate often is <0.01 kPa/h (0.075 mmHg/h) in Atlas aluminum laboratory-scale hypobaric chambers.

Lowering the pressure from 760 to 10 mmHg causes a 76-fold increase in the rate at which water vapor diffuses out from a plant commodity through its stomata, lenticles, intercellular system, and barrier air layers (Figure 2.2). Therefore, an 85% RH at atmospheric pressure and 99.8% RH at 10 mmHg should cause the same rate of commodity water loss. At a moderate humidity and atmospheric pressure, plant matter's temperature stabilizes close to the surrounding air's temperature because the combined heat-transfer capacity of radiation (Section 4.2) and convection (Sections 4.1) is much greater than that of evaporative cooling (Section 4.3). The temperature of lettuce stored in air at 80% RH, 20°C, and 1 atm pressure remained at 20°C,

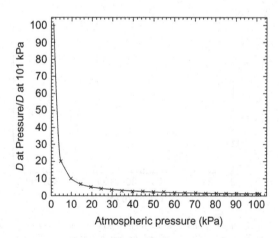

Figure 2.2 Effect of pressure on the air/water-vapor binary diffusion coefficient (D). 1 kPa = 7.5 mmHg. Original figure in public domain: *Reproduced from Figure 1.3 (Corey et al., 2002). Plant responses to rarified atmospheres (Corey, 2000).*

but in a hypobaric plant growth chamber operated at 10 kPa (75 mmHg), 20°C, and 70–78% RH, convection was inhibited and evapotranspiration promoted. This decreased the temperature of the lettuce to the chamber air's 14–16°C dew point (Wilkerson et al., 2004). The leaf temperature of radish seedlings decreased within 2 h at 75 mmHg in a plant growth chamber operated at 70% RH (Peterson and Fowler, 2004). In a similar hypobaric growth chamber, heat radiating from the warmer walls and lights to cooler lettuce, radish or *Arabidopsis* seedlings activated genetic drought–stress mechanisms by providing sufficient latent energy to overly enhance evapotranspiration (Paul et al., 2004; Section 4.3). These examples illustrate the importance of keeping the humidity extremely close to saturation during LP storage in order to prevent a temperature gradient from developing between the commodity and storage chamber's walls, through which heat can be continuously transferred to the commodity by radiation and convection, providing additional latent energy to evaporate (transpire) water and increase weight loss.

Mango storage life progressively increased as the pressure in a leak-tight temperature-controlled hypobaric apparatus was decreased. So little metabolic heat was produced at 15 mmHg, the lowest pressure tested, that the rate of weight loss reached a minimum due to the small amount of latent energy available to evaporate water (Burg, 2004). In studies carried out in leaky chambers, mango and avocado water loss decreased at lower pressures, but increased below 50–80 mmHg due to enhanced leakage (Figure 2.1; Apelbaum et al., 1977b,c; Wang, 1991).

2.2 LP AIR CHANGES MUST BE HUMIDIFIED AT THE STORAGE PRESSURE

A study comparing CA and LP green pepper storage illustrates the error created by humidifying LP air changes at 1 atm pressure before the air enters the vacuum chamber (Hughes et al., 1981). Peppers humidified in this manner and kept in LP at 152, 76, or 38 mmHg became "wrinkled and flaccid, showing severe desiccation," losing weight 4–5 times faster than peppers stored in CA. LP was judged to be the least effective method of storing green peppers, and it was concluded that "water loss thus appears to be a major problem in hypobaric storage." Both the LP and CA chambers were humidified by flushing them with 86% RH atmospheric pressure air, which expanded and dried during entry into the LP chambers. Storage flowing 86% RH air through CA chambers had been compared to passing 4−17% RH rarified air changes through LP chambers. Green peppers are preserved for many months with little weight loss in leak-tight hypobaric chambers when the incoming air changes are humidified after the pressure has been reduced (Burg, 2004, 2005). Hughes et al. claimed that "any improvement in humidification is difficult because of the reduced partial pressure of water vapor." This problem could have been easily rectified by humidifying the air at the storage pressure and temperature, as specified in the original disclosure of the hypobaric method (Burg and Burg, 1966; Burg, 1967). Hughes et al. (1981) refer to studies carried out in 1966 by Tucker at the National Vegetable Research Station in England, which showed desiccation to be the "main problem" in hypobaric storage. Was this due to the same error? Many LP publications do not indicate the pressure at which the air change was humidified because the author(s) are unaware that this is important.

The experimental error caused by humidifying air at 1 atm pressure is also evident in LP mango experiments performed by Ilangantileke (1989), and in tests with potatoes, apricots, peaches, sweet cherries, apples, bananas, avocados, mangos, limes, guavas, pears, and tomatoes performed by Wu and Salunkhe (1972a,b), Salunkhe and Wu (1973), Wu et al. (1972), and Jadhav et al. (1973). Even though mango transpiration raised the measured humidity in Ilangantileke's chambers to a much higher value than saturated 1 atm air could possibly have created after expanding and drying during entry, nevertheless at 60 mmHg, the fruits experienced a 17.2% weight loss in 17 days (Table 2.1). In a leak-free LP chamber humidified to 99.5% RH at 13°C, 15 mmHg (0.1% [O_2]), mangos only lost 3% of their weight during a 56-day storage (Davenport et al., 2006). The weight of Rabbiteye blueberries decreased by 23.04% in LP and 0.99% in NA when both

Table 2.1 "O Krong" Mango Weight Loss During a 17-Day Storage at 13°C and Various Pressures. Chambers Were Humidified with Saturated 1 atm Air				
Chamber pressure (mmHg)	760	150	100	60
% RH expected in empty chamber	100	20	15	8
% RH measured with mangos present	100	85	55	34
Mango weight loss (%) in 17 days	1.5	7.8	13	17.2
Computed from Data Published in Ilangantileke (1989).				

systems were flushed with humidified atmospheric pressure air during a 28-day storage (Al-Qurashi et al., 2005).

Wu and Salunkhe, Salunkhe and Wu, Wu et al., and Jarad et al. inadvertently partly offset their mistake of humidifying incoming air at atmospheric pressure by the additional error of providing inadequate airflow, apparently only 0.048 air changes per hour through each chamber. They were unaware that commodity transpiration and the nearly static airflow restored the humidity to 90−95%, the value they measured in each LP chamber (see Chapter 6). The air-change rate was so low that stored tomatoes consumed essentially all available $[O_2]$ at 102 mmHg, nearly all of the $[O_2]$ at 278 mmHg, and a significant amount at 471 mmHg. Pressures of 278−471 mmHg do not affect ripening when sufficient air changes are continuously passed through a hypobaric storage chamber (Burg, 2004), but in this study the lowered $[O_2]$ partial pressure delayed ripening at 471 and 278 mmHg, compared to the 646 mmHg control (the atmospheric pressure in Logan, Utah, where the experiment was performed). At 107 mmHg, the "anaerobic" tomatoes never ripened and eventually deteriorated.

2.3 A COLD SPOT ON A VACUUM CHAMBER'S SURFACE INCREASES COMMODITY WATER LOSS

Laboratory LP chambers are installed in temperature-controlled rooms or enclosures which may have a nonuniform air distribution pattern that produces a "cold spot" on the vacuum chamber's surface. This makes the LP apparatus susceptible to establishment of an evaporation−condensation cycle between the commodity and chamber cold spot. In a mixture containing water vapor and a noncondensable gas such as air, if the water vapor condenses on a cold surface, the noncondensable air is left behind and the incoming condensable vapor must diffuse through the vapor−gas mixture collected in the vicinity of the condensate surface before reaching and condensing in the stagnant film layer at the cold surface. The presence of noncondensable air adjacent to the condensate surface acts as a thermal resistance barrier to convective heat transfer and reduces the heat-transfer coefficient for condensation by at least an order of magnitude (Section 4.1; Özisik, 1985; Knudsen et al., 1984). A low pressure also promotes water vapor diffusion from plant matter to the cold surface (Figure 2.2; Eq. (3.1)), and increased evapotranspiration lowers the commodity's temperature, causing heat to radiate or transfer by convection from the chamber walls to the cooler commodity, providing extra latent energy to fuel evaporation. Excess weight loss is also likely to occur when plant matter is stored in laboratory systems located within cold rooms in which refrigeration is controlled by an On/ Off cycle with a large dead band, rather than by a continuously operating modulating thermostat, and also if the refrigeration system has defrost cycles.

Condensation has sometimes been noted on a restricted cold zone of a laboratory hypobaric chamber's wall (Lougheed et al., 1978). This promotes commodity water loss because the dew point in the chamber approaches the temperature of the cold surface that is causing condensation to occur. The temperature must be kept as uniform as possible over the entire surface to prevent development of an evaporation/ condensation cycle. In LP intermodal containers, a surface temperature uniformity of $\pm 0.2°C$ is maintained by a jacketed refrigeration system utilizing a secondary coolant (Sharp, 1979). Laboratory systems attempt to provide a uniform "jacket" by blowing constant temperature refrigerated air over the entire surface of the LP chamber (Burg, 2004).

2.4 LP PREVENTS C_2H_4, CO_2, AND NH_3 RETENTION

Regardless of whether tomatoes were stored at atmospheric pressure in 20% [O_2]) or in pure [O_2] at 1/5 atm pressure (20% [O_2]), they ripened at the same rate (Stenvers and Bruinsma, 1975). This experiment, ostensibly demonstrating that lowering the pressure does not remove ethylene and delay ripening, was carried out in sealed 2-L vacuum desiccators. The misleading result occurred because even with permanganate present in the sealed desiccators, sufficient ethylene accumulated each day to maximally stimulate tomato ripening before the desiccator was vented, ventilated, and the storage pressure reinstated (Burg, 2004). Stenvars and Bruinsma erroneously assumed that the threshold for ethylene action was $>5\,\mu L/L$ [C_2H_4], whereas it is only $0.005\,\mu L/L$ in tomatoes (Wills et al., 2001; Burg and Burg, 1965). Banana ripening is delayed by 1.9-fold when pure O_2 is continuously flowed through an LP chamber at 1/5 atm pressure (Burg, 1967).

Tolle (1969) and Goldschmidt et al. (1993) claimed that hypobaric storage cannot displace "'active-bound [C_2H_4]" from within plant tissues, and only influences ripening or senescence by lowering [O_2]. This opinion was disproved by studies showing that in various commodities ethylene's measured dissociation constant from its receptor site has the same value as the applied $[C_2H_4]$ concentration causing a half-maximal biological response (Sisler, 1979). Warner and Leopold (1971) proved that "bound" and intercellular C_2H_4 rapidly equilibrate by demonstrating that recovery from C_2H_4 treatment commenced within 15 min after the elongation growth of etiolated pea seedlings had been inhibited by applied ethylene and the plants were transferred to fresh atmospheric air. Etiolated pea seedlings were selected for this experiment because they do not produce autocatalytic ethylene. Therefore, ethylene binding should be reversed and the gas's action cease after the applied ethylene escaped from the tissue.

Low-pressure storage also reduces intercellular and intracellular $[CO_2]$ and $[NH_3]$. Contrary to the claim that LP lacks the benefit that added CO_2 provides to CA and MA (Lougheed et al., 1977, 1978; Anon, 1974; Abeles et al., 1992), LP's ability to decrease $[CO_2]$ both in the storage atmosphere and within plant matter provides benefits that CA and MA cannot duplicate. CO_2 removal can inhibit mold and bacterial growth (Section 12.5), it prevents CO_2 injury, opens stomata in darkness (Section 3.3), and causes a tissue's ascorbic acid to be retained (Burg, 2004). Excess succinate is toxic to fruits and vegetables and accumulates in them when added CO_2 injures apples (Hulme, 1956; Ransom, 1953; Shipway et al., 1973), pears (Williams and Patterson, 1964; Frenkel and Patterson, 1969; Ulrich, 1975); sweet cherries (Singh et al., 1970); spinach (Murata and Ueda, 1967); citrus (Davis et al., 1973); peas (Wager, 1973); grapes (Ulrich, 1970); apricots and peaches (Wankier et al., 1970). Both *in vivo* and *in vitro*, low $[CO_2]$ stimulates ACC oxidase by as much as 40-fold compared to ethylene forming enzyme (EFE) activity in atmospheric air (Kao and Yang, 1982; McRae et al., 1983; Dilley et al., 1993; Fernández-Maculet et al., 1993; Poneleit and Dilley, 1993; Smith and John, 1993a,b; Yang et al., 1993). CO_2 increases the V_{max} of EFE by 10-fold, the apparent K_m for ACC by several-fold, and EFE's apparent K_m for O_2. The CO_2 concentration required for half-maximal activity is 0.68% (Dilley et al., 1993; Poneleit and Dilley, 1993). LP inhibits ethylene production by creating a nearly CO_2- and O_2-free atmosphere within the commodity, whereas NA and CA storage elevates the ICC sufficiently to support a maximum ethylene production rate. Bluing of red roses is caused by senescence-linked proteolysis of proteins, yielding amino acids and amides that are deaminated to ammonia. The ammonia increases the pH of the cytoplasm and causes a red rose's pH sensitive anthocyanin pigment to turn blue. Because LP delays senescence and promotes diffusive removal of NH_3 (Figure 2.2), it prevents red roses from increasing in cytoplasmic pH and "bluing" during long-term hypobaric storage (Asen et al., 1971; Burg, 2004).

2.5 LP DOES NOT "OUTGAS OR EVACUATE" FLAVOR AND AROMA VOLATILES OR WATER VAPOR

Breaker and light-pink tomatoes do not develop normal flavor and aroma if they reach a red-ripe stage during LP storage at 180–190 mmHg (Tolle, 1972). Likewise, apples fail to develop normal flavor, aroma, and texture during prolonged LP storage (Bangerth, 1984), and the lower the storage pressure, the more quickly this symptom develops and the fruits lose their ability to produce or respond to ethylene. Wu et al. (1972) reported that there is an inverse relationship between the hypobaric storage pressure and quantity of volatiles that can be recovered from tomatoes stored in LP at $12.8°C$. These results prompted researchers to claim that a vacuum "outgases and evacuates" flavor and aroma volatiles (Wu et al., 1972; Salunkhe and Wu, 1975; Tolle, 1972; Lougheed et al., 1977; Kader, 1975), but when the same symptom was observed to arise during apple CA storage, and more rapidly at a lower $[O_2]$ concentration, it became obvious that the effect on apples was caused by hypoxia rather than hypobaria (Shatat et al., 1978; Bangerth, 1984). This likely is also true for tomatoes since the abnormally low air-change rate in Wu et al.'s apparatus caused so much of the chamber $[O_2]$ to be consumed by the fruits that eventually they were exposed to anerobiosis at 102 mmHg, $12.8°C$, and to hypoxia at 278 and even 471 mmHg. None of the volatiles measured in Wu et al.'s study boil at 102 mmHg, $12.8°C$, except acetaldehyde at and below 565 mm Hg (Table 9.1). Therefore, the >95% loss of 11 volatiles in addition to acetaldehyde that occurred at 102 mmHg was not due to boiling. Even the loss of acetaldehyde could not have been due to boiling because the pressure within turgid plant cells is between 4 and 20 atm. When mangos were stored at $13°C$ with 0.075% $[O_2]$ flowing through an LP chamber at 15 mmHg, and 0.14% $[O_2]$ flowing in CA, high concentrations of ethanol and acetaldehyde accumulated in the fruits (Table 8.2). Spalding et al. (1978) found that ethanol accumulated at essentially the same rate in "Iobelle" sweet corn kernels during storage at 50 mmHg (1.4% [O_2]) and in 2% [O_2] CA. Ultralow [O_2] reduces volatile production by Fuji apples, except for ethanol and acetaldehyde (Argenta et al., 2004). This would explain the loss of volatiles in Wu et al.'s experiment (Figure 2.3).

The ethylene sensitivity of preclimacteric bananas did not decrease during 4 months at 40–67 mmHg (0.8–1.0% [O_2], they retained their green color, and after removal from LP storage ripened with normal texture, sweetness, flavor, aroma, and yellow color (Apelbaum et al., 1977a). Bananas stored in 1% [O_2] at atmospheric pressure suffer low-[O_2] damage, within 11 days, lose their ability to produce ethylene, and subsequently cannot ripen with acceptable flavor, sweetness, and color (Hatton et al., 1975; Hatton and Spalding, 1990; Burg, 1990; Apelbaum et al., 1977a). Bananas stored for substantially longer than 4 months in LP behave in the same manner as apples, eventually losing their ability to produce or respond to ethylene and ripen normally (Bangerth et al., 2012). Green to half-yellow papayas, which can only be kept in CA for up to 12 days, do not advance in color during 4 weeks in LP at 10°C, 20–25 mmHg (0.3–0.43% [O_2]), and after transfer to air they ripen with full flavor, aroma, texture, and yellow color. Mangos develop excellent flavor, aroma, and color when they ripen after removal from a 56-day storage in LP at 13°C and 15 mmHg (0.1% [O_2]). Flavor and aroma are also normal after LP storage of loquats, avocados, limes, oranges, guavas, blueberries, strawberries, cantaloupes, carambolas, cherries, asparagus, pineapples, green beans, corn, cucumbers, mushrooms, green peppers, and currants (Burg, 2004). Fruits typically produce only small amounts of flavor and aroma volatiles before ripening commences. Any reduction in the production or content of these substances during LP storage is reversed within a few days after the commodity is transferred to air at atmospheric pressure. Subsequently, flavor and aroma volatiles are produced and accumulate at a normal rate as ripening progresses (Burg, 2004).

2.6 SLOW EVACUATION AND VENTING DO NOT DAMAGE COMMODITIES

An LP warehouse or intermodal container can be rapidly brought to an optimal pressure for storage, quickly vented and opened, and then restarted. This allows commodities to be added or removed at frequent intervals from hypobaric warehouses, and hypobaric intermodal

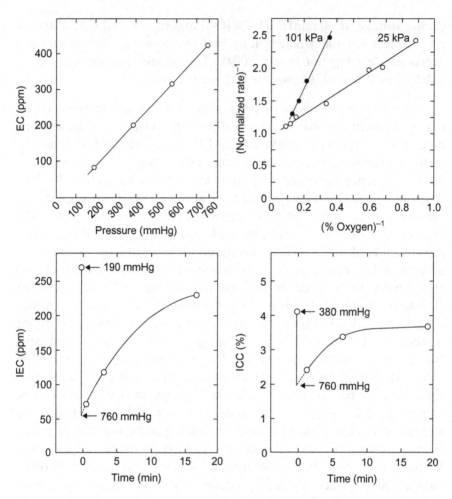

Figure 2.3 Upper left: Equilibrated internal [C₂H₄] concentration (IEC) within McIntosh apples at various hypo-baric pressures. The rate of C₂H₄ production was not reduced at 190 mmHg. Drawn from data in Table 5.2 (Burg and Burg, 1965) entitled "Partial pressure of ethylene within a McIntosh apple at different atmospheric pressures"—by STM permission from Wiley-Blackwell to Elsevier. Upper right: Double reciprocal plots of respi-ratory isotherms for O₂ depletion by Hass avocado fruits at 25°C and either 101 kPa (760 mmHg) or 25 kPa (188 mmHg). The apparent Michaelis Menton constant (K_{m,O2}) is 4.2-fold lower at 25 kPa than it is at 101 kPa due to the 4.04-fold increase in the O₂ diffusion rate at the lower pressure (calculated from Figure 2.2; Eq. (3.1)); Tucker and Laties, 1985; Burg, 2004). Lower: Effect of pressure on the [C₂H₄] (left) and [CO₂] (right) concentration in the intercellular spaces of McIntosh apples. The IEC and internal [CO₂] concentration (ICC) were determined at 760 mmHg, and then the fruit was placed in LP at 190 mmHg (left) or 380 mmHg (right) for 120 min. An arrow indicates the time at which the vacuum was released. The subsequent curve shows the recovery of the IEC and ICC at 1 atm. Extrapolating the curves to the moment when the vacuum was released indicates the IEC and ICC during LP storage. Reproduced from Burg and Burg (1965), Gas exchange in fruit. Figure 2.1. Effect of atmospheric pressure on the ethylene (to the left) and CO₂ (to the right) content of McIntosh apples. STM permission from Wiley-Blackman to Elsevier.

containers to be opened for inspection and then restarted at a port-of-entry. CA warehouses and intermodal containers do not provide these conveniences. Chrysanthemum cuttings have been returned to atmospheric pressure for up to 8 h each day without a significant diminution in their LP storage life (Burg, 2004). Roses retained nearly their entire initial vase-life during 7 weeks in an LP warehouse (Figure 10.3), even though they were stored in boxes that absorbed water (see Chapter 7) and each week the warehouse was vented and remained at atmospheric pressure for 2–4 h while inventory was removed or added, after which the warehouse was again evacuated and the cycle repeated (Figure 10.2). Roses can only be kept in NA or CA for at most 5–10 days, after which they "blow open" and display only a few days' vase-life. Bananas kept in a hypobaric atmosphere for 3 h daily, and then returned to atmospheric pressure each day, did not develop a climacteric and remained green during a 15-day experiment (Awad et al., 1975).

Lougheed et al. (1978) suggested that pump-down and venting might damage plant matter. The magnitude of the pressure gradient that develops across a commodity's surface during pump-down or venting depends on the speed at which these processes are carried out since air tends to equilibrate across the plant matter's induced surface pressure gradient at a finite rate. The structure of most horticultural commodities is better suited to withstand decompression than compression, and is not damaged or deformed by the limited compressive force that develops in LP warehouses and intermodal containers during pump-down (Wheeler et al., 2011; Burg, 2004). Even a 10- to 15-min pump-down in a laboratory hypobaric apparatus or during commercial vacuum cooling does not cause injury. Most horticultural commodities are not damaged by rapidly venting a hypobaric storage apparatus or a commercial vacuum cooler, but there are exceptions. Grapefruits become disfigured, and strawberries, other berries, cherries, and more than half-yellow papayas may exude juice and discolor if the venting is too rapid (Burg, 2004). Lagunas-Solar et al. (2006) reported that 10 rapid decompression (evacuation) and compression (venting) cycles during a 10- to 15-min interval caused textural damage (juicing) of raspberries and blackberries, discoloration of lettuce, strawberries, and blackberries, but no obvious damage to oranges, lemons, grapes, blueberries, and bananas. Sensitive commodities are not injured by venting if the compressive force that develops is minimized by prolonging the repressurization so that it requires at least 30 min (Burg, 2004).

CHAPTER 3

Gas and Vapor Mass Transfer at a Low Pressure

3.1 DIFFUSION OF GASES AND VAPORS

Diffusion is the random thermal motion of molecules leading to their net transfer from a region of higher concentration to a region of lower concentration. Outward water vapor diffusion (transpiration) through air-filled stomata, lenticles, and pedicle end scars is the major unidirectional mass-transport process in plant matter. Transpiration accelerates the net outward flux of CO_2, NH_3, and VOCs such as ethanol and acetaldehyde from plant matter, and restricts the net inward flux of O_2. The diffusion coefficient (D) is the ratio between the concentration gradient and the transfer rate of the diffusing gas/vapor mixture across unit area. For a binary air/water−vapor mixture, D can be computed by the Fuller relationship (Fuller et al., 1966):

$$D = \frac{10^{-3} T^{1.75} [(M_A + M_V)/M_A M_V]^{1/2}}{P\left[\left(\sum_v\right)_A^{1/3}\right] + \left[\left(\sum_v\right)_V^{1/3}\right]^2} \tag{3.1}$$

where T is the temperature (K), P is the pressure (atm), M_V and M_A are the molecular weights of water ($M_V = 18$) and air ($M_A = 28$), $(\sum_v)_A$ and $(\sum_v)_V$ are the atomic diffusion volumes of air (20.1) and water−vapor (12.7), and D is the binary diffusion coefficient (cm^2/s). The diffusion rate is inversely proportional to the pressure and directly proportional to $T^{1.75}$ (Figure 2.2; Eq. (3.1)).

Pressure influences the internal concentration of gases and vapors in plant matter through its effect on diffusion. This was confirmed by studies in which the outward diffusion coefficient of ethylene from apples was artificially increased by substituting helium ($M_H = 4$) for atmospheric N_2 ($M_N = 28$) in the ambient air (Burg and Burg, 1965). At 22°C, the binary diffusion coefficient of an 80% [He] + 20% [O_2] mixture is 2.68-fold higher than that of air Eq. (3.1), and apples placed in this mixture rapidly lost 45% of their internal ethylene. In 94% [He] + 6% [O_2] ethylene diffusion is accelerated 3.72-fold, and the IEC

Hypobaric Storage in Food Industry. DOI: http://dx.doi.org/10.1016/B978-0-12-419962-0.00003-6

of apples decreased by 63%. The measured ethylene production rate of the apples was the same in air, 80%[He] + 20% [O_2], and 94% [He] + 6% [O_2].

Gas and vapor diffusion from a commodity's cytosol to the ambient atmosphere can be analyzed using equations governing electric circuitry by assigning a resistance to each barrier encountered en route. Concentration gradient ($c_o - c_i$) is substituted for EMF (V) and gas or vapor flux (J_V) in place of electric current (I). Then, according to Fick's law and by analogy to Ohm's law ($V = IR$):

$$(c_o - c_i) = J_v r \tag{3.2}$$

where J_v (cm^3/cm$^2 \cdot$ s = cm/s) is the volume flux density across the particular barrier, c_o and c_i are the concentrations (mols/cm^3) outside (o) and inside (i) the barrier, and r (atm $\cdot$ s/cm; usually written s/cm) is the gas or vapor's resistance to diffusive transport through the barrier. Just as an electric wire's resistance [R(ohms) = $\rho L / A_{cs}$] depends on the wire's resistivity (ρ, ohm-m), length (L, cm), and cross-sectional area (A_{cs}, cm^2), likewise the resistance value assigned to each diffusion barrier is directly proportional to the thickness of the barrier (Δx, cm), and inversely related to both the surface area (A, m^2) through which diffusion is occurring, and the gas or vapor's diffusion coefficient (D, cm^2/s) in the media comprising the barrier. The units used to indicate a barriers conductance, abbreviated cm/s, refer to the cm^3 of gas transferred each second per cm^2 of surface. The barrier's resistance, abbreviated s/cm, is the reciprocal of its conductance, and like conductance is defined per cm^2 of surface area.

At equilibrium the rate at which each gas or vapor is produced or consumed equals its flux ($A_{com}J_v$) across the commodity's surface area. The same flux must occur through the intercellular system, across the total cellular surface area including stomata, lenticles, and pedicle end scars, through the surface of the box and any protective wraps in which the commodity is transported or stored, and through all associated barrier air layers. To accurately describe the concentration gradients created when an identical flux passes through the various barriers of a resistance network, the resistance value for each barrier is referenced to the commodity's surface area (A_{com}). This correction adjusts for the fact that when a barrier's surface area is greater, a smaller concentration gradient drives the same flux across it. The overall effect is to create an imaginary network in which each barrier is assigned the same surface area, and the resistance values are adjusted to reflect the relative ease of transport through each step in the network. The "effective" resistance value (r_{com}) to be used for each barrier in the resistance network formulation is:

$$r_{com} = (A_{com}/A°)r° \qquad (3.3)$$

where the total surface area of a particular barrier is $A°$, and the barrier's measured resistance is $r°$ per cm^2 of its surface area. An area correction is also applied to the box's resistance in order to reference it to the commodity's total surface area. The effective box resistance (r_b) to be used in a resistance network calculation (Section 3.1) is $r_{box}(\sum A_{com}/A_{box})$ where $\sum A_{com}$ is the total surface area of commodity in the box, and r_{box} is the measured box resistance. These adjustments allow the laws of parallel and series circuits to be used in summating resistances in the network:

$$\begin{aligned} r_1 + r_2 + r_3 = r_{1,2,3} \quad \text{(series resistances)} \\ 1/r_1 + 1/r_2 + 1/r_3 = 1/r_{1,2,3} \quad \text{(parallel resistances)} \end{aligned} \qquad (3.4)$$

The resistances of boxes and wraps, stagnant air layers, the commodity's skin, intercellular air space, cell walls, plasma membrane, cytosol, and membranes of cellular inclusions act in series. The resistances of lenticles, stomata, and the pedicel end scar act in parallel and together constitute the surface's pore resistance, which acts in parallel with the skin's cuticular resistance and in series with other resistances.

To simplify the calculation of gas and vapor movement through a resistance network containing both air and liquid phases, the gas and vapor amounts in each liquid phase are expressed as the equilibrium concentration in an air phase which gives rise to the relevant liquid phase concentration. Gases and vapors will then diffuse through the liquid and air phases toward regions of lower concentration regardless of the partition coefficients and concentrations in the liquid phases involved in the transport. The liquid phase concentration is multiplied by the partition coefficient (K), where K is the concentration of all forms of the gas or vapor in the liquid phase capable of penetrating the barrier, divided by the equilibrium gas or vapor concentration in an adjacent air phase.

The initial route of CO_2 and the final pathway for O_2 is through their concentration gradient in the cytosol. When O_2 consumption occurs uniformly throughout a spherical cell (Goddard, 1947; Burton, 1982):

$$p_o - p_x = \frac{\dot{Q}_{O_2}(R^2 - x^2)}{6D} \tag{3.5}$$

where p_x is the partial pressure of $[O_2]$ (atm) at distance x (cm) from the center of the cell, p_o is the partial pressure of $[O_2]$ (atm) at the cell's periphery; $UNDEFINEDnedot$; Q_{O_2} is the O_2 consumption rate (cm^3/min per cm^3 of tissue); D is the diffusion coefficient of O_2 in the cell sap $(1.4 \times 10^{-5}$ cm^3/cm$^2 \cdot$ min $\cdot$ atm); and R is the radius of a spherical cell (cm). For a 5×10^{-3} cm radius apple cell, $p_o - p_c = 0.29 q_{O_2}$, and at 0°C the $[O_2]$ gradient from the periphery to the center (c) of an apple cell respiring at a typical rate of 4×10^{-5} cm^{3}O$_2$/min per cm^3 of tissue would only be 1.16×10^{-5} atm of equilibrated partial pressure in an air phase. If the cell periphery contained dissolved O_2 equilibrated with 20% $[O_2]$ in an air phase, the center would be in equilibrium with 19.9988% $[O_2]$ and the surface-to-center cellular $[O_2]$ gradient would only be 0.0012%. For the same rate of CO_2 production, the center-to-surface CO_2 gradient would be increased by 0.0012%. The binary diffusion coefficients of various metabolic gases in water are not very different, so they all rapidly diffuse through the cytosol across a very small concentration gradient.

Gases pass through the cell wall diffusing in water present in tortuous interstices approximately 10 nm in diameter, occupying slightly less than half of the cell wall's volume. Typically the cell wall's gas conductance is about threefold larger than the plasma membrane's conductance (Noble, 1991), and the combined conductance of the cell wall, plasma membrane, and cytosol is approximately 5000-fold greater for gases than it is for ethanol, which is one of the most rapidly penetrating VOCs (Collander, 1937).

The rate at which gases and vapors diffuse is inversely related to the pressure (Figure 2.2; Eq. (3.1)). At atmospheric pressure, the extracellular air-filled barriers determine how rapidly gases enter and exit the commodity because the resistance to gas transport through air present in a commodity's intercellular spaces, lenticles, pedicle end scar, stomata, barrier air layer, and in micropores and stagnant air layers associated with storage boxes and wraps, is much larger than the combined cytosol, plasma membrane, and cell wall gas transport resistances. The optimal hypobaric pressure for preserving horticultural commodities is usually between 7.5 mmHg—the absolute pressure at the Martian surface—and 20 mmHg—the pressure in the earth's stratosphere at an altitude of 25 km (15.5 miles). In these low-density environments, LP accelerates the rate of gas and VOC diffusion through air-filled barriers by up to 100-fold (Figure 2.2; Eq. (3.1)), promoting the escape of metabolic CO_2, C_2H_4, NH_3, CN, NO, and VOCs from the commodity, reducing the cellular liquid-phase concentrations of these gases and vapors, and increasing O_2 diffusion into plant matter.

Enhanced gaseous diffusion through air-filled extracellular barriers during LP storage provides benefits that CA, MA, and NA cannot duplicate. By accelerating the diffusive escape of metabolic ethylene, LP prevents ripening and other ethylene-dependent responses. Enhanced diffusive entry of O_2 allows commodities to be stored without injury in an ultralow $[O_2]$ concentration that inhibits the growth and sporulation of many bacteria and fungi. There is so little $[O_2]$ available at 20 mmHg, 10°C, that a papaya's ATP production is inhibited by 95.8% (Figure 8.1, right) compared to its ATP production at atmospheric pressure. The pausity of ATP prevents energy requiring processes such as ripening and senescence from occurring. Mangos are injured by less than 3–5% $[O_2]$ at 1 atm pressure and can only be stored in NA or CA for 14–21 days if the $[O_2]$ concentration is tolerable (Bender et al., 2000), but *fully mature* mangos are preserved for at least 8 weeks in LP at 13°C and 15–20 mmHg (0.08–0.24%$[O_2]$) without low-$[O_2]$ damage (Davenport et al., 2006). Respiratory and fermentative heat provide the latent energy used to evaporate (transpire) water, and because LP markedly inhibits metabolic heat production, the mangos only experience a 3–4% weight loss in 2 months. Bluing of roses is caused by senescence-linked proteolysis of proteins, yielding amino acids and amides that are deaminated to ammonia. Within a few weeks, free ammonia nitrogen increases by 230% in red roses stored at atmospheric pressure, elevating their cellular pH to 4.5 from 4.0, turning the flower's pH-sensitive red anthocyanin pigment blue (Asen et al., 1971). CA does not significantly affect the storage life or delay the bluing of red roses. By delaying senescence and enhancing the diffusive escape of NH_3 gas (Figure 2.2), LP prevents roses from increasing in pH and turning blue.

3.2 CARBON DIOXIDE

Since Kidd and West's pioneering work on CA storage (1927a,b), it has been axiomatic that moderate [CO_2] concentrations often benefit storage life. Therefore, postharvest physiologists have assumed that because LP does not elevate [CO_2], the hypobaric method lacks the advantages that CO_2 affords to CA (Abeles et al., 1992; Lougheed et al., 1977, 1978; Anon, 1974; Kader, 1975). Attempts have been made to add CO_2 during hypobaric storage in order to overcome this inferred deficiency (Spalding and Reeder, 1976a; Haard and Lee, 1982). The effects of abnormally low [CO_2] could not be critically evaluated until LP became available because previously there was no way to continuously remove CO_2 from within a commodity's intercellular spaces. LP's ability to drastically decrease both the ambient and intercellular [CO_2] levels has proven to be an important advantage providing benefits that cannot be duplicated by elevating [CO_2] (Section 2.4). Some of the most notable LP effects resulting from unusually low [CO_2] are stomatal opening in darkness (Kirk et al., 1986; Kirk and Anderson, 1986; Veierskov and Kirk, 1986); depressed growth of bacteria, fungi, and yeasts (Monod, 1942; Rochwell and Highberger, 1927; Valley and Rettger, 1927; Wells, 1974; Wells and Uota, 1970; Apelbaum and Barkai-Golan, 1977; Wu and Salunkhe, 1972a; Haard et al., 1979; Section 12.5); retention of ascorbic acid (Bangerth, 1974, 1977; Dilley, 1977; Izumi et al., 1999; Spalding and Reeder, 1976b); inactivation of ethylene forming enzyme (Yang et al., 1993; Dilley et al., 1993; Smith and John, 1993a,b); limiting toxic succinate accumulation; and preventing [CO_2]-injury (Section 2.4).

3.3 STOMATAL OPENING

Stomata respond to the [CO_2] concentration in the substomatal crypt (Raschke, 1975; Willmer and Fricker, 1996) and are sensitive to [CO_2] in darkness and light to the same degree (Raschke et al., 1970). CO_2-free air opens stomata in darkness (Zelitch, 1969; Willmer and Fricker, 1996; Murray, 1997) and elevating [CO_2] from 0% to 0.035% progressively closes them. High [CO_2] shuts stomata in the light and photosynthesis opens them by depleting [CO_2] in the substomatal crypt (Wilmer and Fricker, 1996; Murray, 1997).

Anaerobiosis prevents stomata from opening in light, closing in darkness, and responding to [CO_2] in light or darkness (Walker and

Zelitch, 1963; Couchat, 1977; Couchat and Lascève, 1980; Couchat et al., 1982; Akita and Moss, 1973; Akita and Mitasaka, 1969), but in 1.0−1.5% [O_2] at atmospheric pressure stomata behave as though they are functioning in normal air (Akita and Mitasaka, 1969; Akita and Moss, 1973; Louguet, 1968, 1972). In darkness, potato, wheat, and soybean stomata open in 0.5−1% [CO_2] + 20% [O_2] (Wheeler et al., 1999; Levine et al., 2009). *Kalanchoe* stomata open fully in 2%[O_2] + 0−15% [CO_2] (Bredmose and Nielsen, 2009). Oat leaf stomata open slightly in 21%[O_2] + 0%[CO_2], wider in 21% [O_2] + 0.035% [CO_2], fully in 21% [O_2] + 5% [CO_2], and they close tightly in 0.5% [O_2] + 0% or 5% [CO_2] (Veierskov and Hansen, 1992).

Reducing the storage pressure lowers the CO_2 and O_2 partial pressure in incoming atmospheric air and the substomatal crypt by causing the air to expand. The resultant low-[O_2] further decreases [CO_2] in the storage chamber and substomatal crypt by inhibiting respiratory CO_2 production. At a low enough pressure [CO_2] in the substomatal crypt should decline sufficiently to open stomata. This expectation has been confirmed by scanning electron micrographs of *Hibiscus* cuttings (Kirk et al., 1986) and microscopic examination of acrylic templates prepared from *Caladium bicolor* and *Colocasia esculentum* (Taro) plants kept in darkness for 5 days at 13.3°C, 20 mmHg; in cucumbers, green beans, and papaya fruits after 2 days in darkness at 10°C, 20 mmHg; and in Valencia orange fruits stored in darkness for 16 days at 4°C or 10°C (Davenport, unpublished)—also by diffusion−resistance measurements in oat-leaf segments (*Avena sativa* L.) indicting full open stomata at 12 mmHg and half-open stomata at 578 mmHg (Veierskov and Kirk, 1986); in cuttings of 13 ornamental species stored for 5 weeks at 15°C, 15 mmHg (Kirk and Andersen, 1986); and in *Hibiscus rosasinensis* L. cv. Moesiana cuttings throughout 8 weeks at 15°C, 15 mmHg (Kirk et al., 1986). In atmospheric air *Hibiscus* stomata closed in darkness within 2 h and remained closed during 8 subsequent weeks. After these cuttings were removed from hypobaric storage their open stomata closed unusually slowly in darkness, did not respond to a low humidity, and needed many hours to regain normal function (Kirk et al., 1986). Introducing 5% [CO_2] while venting the LP chamber did not close these stomata. Instead the CO_2 opened them wider. Likewise, stomata did not close when 5% [CO_2] was introduced during venting of an LP chamber storing oat-leaf segments, and the stomata required many hours to regain normal function at atmospheric pressure (Veierskov and Kirk, 1986). The failure of 5% [CO_2] to close stomata after LP storage agrees with studies demonstrating that 0.5–1% [CO_2] opens stomata in darkness at atmospheric pressure (Wheeler et al., 1999; Levine et al., 2009). Stomata in potted *Caladium bicolor* and *Colocasia esculentum* plants remained open for a prolonged time in dim light after the plants were removed from LP storage and caused the plants to wilt even though the soil was watered (Davenport, unpublished).

Air does not need to be free of all CO_2 to stimulate stomata to open. *Pelargonium zonale* stomata closed in darkness in atmospheric pressure air containing 374 μL/L [CO_2], and opened when [CO_2] was decreased to 200 μL/L, the concentration present in air at 434 mmHg (Louguet, referred to in Zelitch, 1969). Within 24 h in darkness, oat-leaf stomata were nearly fully open in 578 mmHg air containing 266 μL/L [CO_2] (Veierskov and Kirk, 1986).

Increasing the temperature from 0°C to 25°C opens stomata wider (Feller, 2006; Willmer and Mansfield, 1970; Soni, 2010; Rodriguez and Davies, 1982; Serna, 2006; Pemadasa, 1977) because water expands (water's cubical expansion coefficient = 0.00215/°C) and metabolic ion uptake and osmotic water entry into guard cells increases. Commodities warm during air transport[1] and this may explain why stomata present in 4°C cucumbers opened during a 6-h air-transport simulation at an air pressure of 533 mmHg and 20°C temperature, a condition typical of aircraft cargo holds (Laurin et al., 2005, 2006). More than 1.2% respiratory [CO_2] should have accumulated in sealed desiccators during the simulated 6-h cucumber air transport (Eaks and Morris, 1956), and in darkness stomata are opened by 0.5% [CO_2] (Wheeler et al., 1999; Levine et al., 2009). This does not explain why a 6-h exposure to 535 mmHg delayed closure of cucumber stomata for several days after they were returned to atmospheric pressure in darkness, because essentially this same closure response occurred after cucumbers were ventilated rapidly enough during the air-transport simulation to prevent any significant accumulation of CO_2 in the desiccators. Laurin et al. (2005, 2006) attributed this result to "low-pressure

[1]The flower temperature increases to at least 7.2–12.2°C during a 3.5 h air shipment of 1–2°C flowers from Bogota, Colombia, to Miami, Florida, induced in cucumbers by a 6-h exposure to 533 mmHg did not occur during an 8-h exposure of strawberries to 533 mmHg (Laurin et al., 2007) because strawberries are an aggregate fruit lacking functional stomata.

stress," but plants grow and their stomata function normally at 10,000 ft altitude (533 mmHg). Zapotoczny et al. (2003) found that during a 30-day test comparing cucumber storage at 14°C, 98% RH, and either 75 or 760 mmHg, there was no significant difference in weight loss and no evidence of "low-pressure stress." The transient opening of stomata and associated weight loss induced in cucumbers by a 6-hour exposure at 533 mmHg (Laurin et al., 2003) did not occur because strawberries are an aggregate fruit lacking functional stomata.

During both LP storage and air transport, the mechanism controlling stomatal opening is altered within 1 day in response to a temporary decrease in pressure and $[O_2]$. It reverts to its initial state within a few days after the pressure and $[O_2]$ are restored to their initial values (Laurin et al., 2005, 2006). The timing of this behavior resembles that for induction of the gene expression changes that promote fermentative metabolism when the pressure is lowered, and the recovery to normalcy that occurs when the pressure is restored to one atmosphere. These changes originally were attributed to "low pressure stress" until it became evident that they are caused by low-$[O_2]$ rather than a low pressure (see Chapter 8). Likewise, Brisker (1980) and Goldschmidt et al. (1993) claimed that "low pressure stress" induced "wound" ethylene production when grapefruits were stored in LP. Instead, ethylene production increased because hypobaric storage lowered endogenous $[C_2H_4]$ sufficiently to overcome an auto-inhibition of ethylene biosynthesis (Burg, 2004). There is no data indicating that LP causes "low-pressure stress."

Changes in gene expression are involved in the control of stomatal aperture (Sirichandra et al., 2009; Vahisalu et al., 2010), and likely account for the protracted opening of stomata during and after air transport or hypobaric storage. To prevent extra water loss after stomata-bearing tissue is removed from LP storage, it can be kept protected in its water-retentive packaging until its original genetic state has been restored.

3.4 VOLUMETRIC EXPANSION

Expansion dries and purifies incoming atmospheric air as it enters a hypobaric system. This lowers the partial pressures of $[O_2]$, $[CO_2]$, $[C_2H_4]$, other gases, VOCs, water vapor, and cloud condensation nuclei (CCNs) present in the storage atmosphere in proportion to the pressure reduction. Heterogeneous condensation of super-saturated water vapor produced by a combination of mechanical humidification and commodity transpiration (see Chapter 7) is unlikely to occur because expansion of the incoming air reduces the concentration of CNNs by 140-fold at 10 mmHg, 0°C, and 86-fold at 20 mmHg, 13°C.

3.5 AIR CHANGES

At a hypobaric pressure, fewer kilograms of air have to be passed through a chamber to affect an air change, and only a small amount of sensible heat is admitted. When the expanded low-density air is cooled to 0°C, it does not reach its dew point. Therefore, no latent heat needs to be removed. The air-change rate in a conventional refrigerated intermodal container is limited to one per hour to avoid exceeding the refrigeration system's capacity, but in LP a much higher air-change rate does not create a significant refrigeration load (see Chapter 13).

Commercial CA systems are sealed and operate without air changes. In contrast, LP is a flow-through method that replaces $[O_2]$ consumed by respiration while simultaneously preventing metabolic C_2H_4, CO_2, and NH_3 from accumulating within or around the commodity (Section 3.5). Consequently mixed loads of ethylene-producing and ethylene-sensitive plant matter, which cannot be stored together in CA or NA, can be combined in LP (Burg, 2004). Both CA and LP lower the $[O_2]$-partial pressure, but even in that respect the methods differ significantly since optimal LP storage occurs at 1/3 to 1/40th the O_2 partial pressure that causes low-O_2 damage to plant matter in CA (Table 8.1).

To prevent biosynthesized ethylene from accumulating and shortening storage life, warehouses and intermodal containers sometimes are equipped with ventilation systems, ethylene-absorbing filters, or ethylene-destroying means. Ideally, the storage air should contain less than ethylene's 5−10 ppb threshold for a biological response in plant matter (Wills et al., 2001) but no commercially available ethylene-destroying device is that effective. LP purifies incoming air by expansion (Section 3.4), ethylene present in the storage air is removed by continuous air changes (Section 3.5), a low-pressure diminishes the commodity's cellular ethylene concentration and IEC due to enhanced gaseous diffusion from the commodity's intercellular system to the ambient atmosphere (Figures 2.2 and 2.3), and so little $[O_2]$ and $[CO_2]$ is present in the rarified air that ethylene production is inhibited by an additional 90% (Section 2.4). Ethylene accumulates at atmospheric pressure because more ethylene is produced and the ventilation system of a conventional (NA) refrigerated intermodal container is restricted to less than a few air changes per hour to avoid exceeding the refrigeration system's capacity.

CHAPTER 4

Heat Transfer at a Low Pressure

Worldwide, nearly one-third of horticultural commodities spoil during distribution (Kader and Rolle, 2004; N.A.S., 1978), 5–25% in developed countries and 20–50% in developing countries (Kader, 2002; Saraswathy et al., 2010; U.N. Food and Agricultural Organization—FAO). Poor humidity control and an elevated temperature are the major causes of 25–35% spoilage in aircraft holds (Laurin et al., 2006; Agrotechnology and Food Science Group, Wageningen). The RH in NA and CA warehouses and intermodal sea containers is 90–95% and the temperature is controlled ±1.5°C. In LP intermodal containers and warehouses, the RH exceeds 99% and the temperature varies by ±0.2°C (Sharp, 1985; Burg, 2004).

When there is a temperature difference, heat flows from regions of high to low temperature, and once the temperature distribution is known, the rates of heat transfer can be determined from laws relating heat flux to the temperature gradient. Conduction, convection, radiation, and evaporation (or condensation) modulate the vapor-pressure and temperature gradients that develop in systems containing biological material, and heat transferred by evaporative cooling determines the rate of water loss. During hypobaric storage, most of the respiratory (Eq. (4.16)) and fermentative (Eq. (4.17)) heat is transferred by the major available heat transfer mode, evaporative cooling, and the commodity must be kept warmer than the storage air and chamber wall to prevent radiation and convection from providing it with environmental heat that would evaporate additional water and increase weight loss.

4.1 CONVECTION

Forced convective heat transfer (Q_c, watts) to or from an object is given by:

$$Q_c = h_m A(\Delta T) \tag{4.1}$$

Hypobaric Storage in Food Industry. DOI: http://dx.doi.org/10.1016/B978-0-12-419962-0.00004-8

where h_m is the mean convective heat transfer coefficient (W/m² · °C), A is the object's surface area (m²), and ΔT the temperature difference (°C) between the object and ambient air. The mean film coefficient for forced convection is (Knudsen et al., 1984):

$$h_m = 0.678k(Re_L)^{1/2}(Pr)^{1/3}L^{-1} \tag{4.2}$$

where k is the thermal conductivity (W/m · K), L is a characteristic dimension (m), and the Reynolds ($Re_L = Lv\rho/\mu$) and Prandtl ($Pr = c_P\mu/k$) numbers[1] are dimensionless values used to simplify engineering calculations, ρ is the gas mixture's density (kg/m³), v the velocity of flow (m/s), μ the dynamic viscosity (kg/m · s), and c_P the specific heat at constant pressure (kJ/kg · K).

The water vapor mole fraction in saturated air at 1 atm is 0.006 at 13°C and 0.012 at 0°C; it is 0.56 in LP at 20 mmHg, 13°C, and 0.46 at 10 mmHg, 0°C. The higher water–vapor mole fraction in LP lowers the air/water–vapor mixture's specific heat and increases its dynamic viscosity and thermal conductivity. The mixture's density (ρ) is inversely proportional to the pressure. These changes alter Re_L and Pr during hypobaric storage, reducing h_m at 10 mmHg to approximately 11% of its value at 1 bar (Burg and Kosson, 1982, 1983; Eqs. (4.1) and (4.2)). The heat transfer coefficient for forced convection is independent of the temperature gradient, but the free-convective film coefficient depends on the temperature difference between air and the surface being cooled since this creates the required buoyant flow (Eq. (4.4)). For a perfect gas, the buoyancy force (β) for free convection is:

$$\beta = 1/T \tag{4.3}$$

where T is the temperature (K). Because in biological systems the commodity transfers heat by evapotranspiration, there is an additional buoyancy term due to the lower molecular weight of the evaporated water vapor ($M_V = 18$) compared to air ($M_A = 28.9$). At atmospheric pressure and a 0–15°C storage temperature the buoyancy caused by water vapor's low molecular weight is small enough to be disregarded because water vapor represents only a 0.6–1.5% mole fraction of the air/water–vapor mixture. During LP storage, water vapor's buoyancy

---

[1] Pr = (momentum diffusivity)/(thermal diffusivity) = 0.71 for air in the physiological temperature range at 1 bar. Flow is laminar at Re <3000, turbulent at Re >3000.

is highly significant because water vapor constitutes upward of 50% of the gas–vapor mixture. The combined expression for the buoyancy term β during LP storage can be written as:

$$\beta = (1/T)(1 + \alpha) \qquad (4.4)$$

Burg and Kosson (1982, 1983) described the evaluation of α from a fruit to air within a box, and from a box to air in a vacuum chamber.

Gas and vapor density, chemical potential, and heat capacity decrease when the pressure is lowered (Burg, 2004). The film coefficient for natural convection is proportional to the thermal conductivity of the atmosphere and the square root of the pressure. In addition natural convection can be limited by the restricted heat-carrying capacity of a low-density air/water–vapor mixture. At 20 mmHg and higher pressures, free convection is limited by the heat transfer coefficient rather than heat-carrying capacity; at 10 mmHg it is limited to some extent by heat-carrying capacity; at 4.6 mmHg, the pressure used for LP meat, poultry, fish and shrimp storage, the heat-carrying capacity effect predominates (Burg and Kosson, 1982, 1983).

Condensation occurs when water vapor contacts a surface that is colder than water's saturation temperature. The condensed liquid water flows over the surface under the action of gravity, creating a smooth film whose thermal resistance to heat flow determines the convective coefficient for film-wise condensation. If only water vapor and no air is present, the heat transfer coefficient for condensation varies from 4000 to 11,000 W/m$^2 \cdot$ °C on vertical surfaces, and between 9000 and 25,000 W/m$^2 \cdot$ °C on horizontal tubes, compared to 5 W/m^2 for free convection in air through a 25°C temperature gradient to a 0.25 m vertical plate. In LP, the small amount of noncondensable air that is mixed with water–vapor tends to accumulate at the condensate surface when condensation occurs. The buildup of noncondensable air near the condensate film decreases the heat transfer coefficient and rate of mass and energy transfer by inhibiting the diffusion of water vapor from the bulk mixture to the liquid film. The low laminar velocity over the condensate surface, characteristic of an LP system, increases the accumulation of noncondensable gas near the surface and reduces the convective coefficient for film-wise condensation. Simultaneously, the low storage pressure accelerates the diffusion of water vapor through the noncondensable air accumulating near the surface (Figure 2.2; Eq. (3.1)), and limits the adverse effect this air has on the heat transfer coefficient for condensation (Özisik, 1985; Kreith and Bohn, 1997).

4.2 RADIATION

Radiant-heat transfer through a nonparticipating media separating the surfaces of objects occurs independent of pressure. A portion of the total radiation incident on a surface is absorbed by the material, a part is reflected from the surface, and the remainder is transmitted through the body. The *absorptivity* (α) of a surface is the fraction of the total irradiation absorbed by the body; *reflectivity* (ρ) is the fraction reflected; *transmissivity* (τ) is the fraction transmitted. If an energy balance is made on a surface involving radiation as the only heat-transfer mode:

$$\alpha G + \rho G + \tau G = G \tag{4.5}$$

where G is the rate at which irradiation is incident on the surface. According to Eq. (4.5), the sum of absorptivity, reflectivity, and transmissivity must equal unity:

$$\alpha + \rho + \tau = 1 \tag{4.6}$$

Bodies such as boxes and plant or animal matter are opaque and do not transmit incident irradiation. For opaque bodies, $\tau = 0$ and Eq. (4.6) reduces to:

$$\alpha + \rho = 1 \tag{4.7}$$

If the surface of an opaque body is also a perfect reflector from which all irradiation is reflected, $\rho = 1$ and both the transmissivity and absorptivity equal zero. In the physiological temperature range, the polished "shiny" aluminized side of Mylar® film has a reflectivity of $\rho \cong 0.96$ and an emissivity (ε) of 0.04 (Table 4.1). Therefore, Mylar® serves as a highly effective radiation shield.

The emissivity of a surface is the ratio of the energy emitted by a real surface to that of an equally sized and shaped blackbody emitting at the same temperature. Emissivity is always between zero and unity since a blackbody emits the maximum possible radiation at a given temperature. Graybodies are surfaces with monochromatic emissivities and absorptivities whose values are independent of wavelength. The

Table 4.1 Emissivity of Opaque Surfaces in the Physiological Temperature Range	
Material	Emissivity
Aluminum	
Bright foil	0.04
Oxidized	0.20
Cardboard	0.81
Painted (white)	0.05
Water or ice	0.97
Green plant leaf	0.96
Data from www.icess.ucsb.edu/modis/EMIS/leaf.htme; Özisik (1985) Knudsen et al. (1984).	

graybody assumption that the spectral hemispherical emissivity (ε), spectral hemispherical absorptivity (α_λ), and spectral hemispherical reflectivity (ρ_λ) are uniform over the entire wavelength spectrum simplifies the analysis of radiative heat transfer since then, according to the Kirchoff law, $\alpha = \varepsilon$.

In a container filled with cargo boxes, the total box surface area facing the wall (A_1) is the same as the wall surface area facing the box (A_2), and net radiation from A_1 to A_2 (Q_{1-2}, watts) equals the radiation energy leaving A_1 that strikes A_2 minus the radiation energy leaving A_2 that strikes A_1:

$$Q_{1-2} = A\sigma(T_1^4 - T_2^4)/[(1/\varepsilon_1) + (1/\varepsilon_2) - 1) + 1] \qquad (4.8)$$

where ε_1 and ε_2 are the emissivities (Table 4.2) and T_1 and T_2 the temperatures (K) of the surfaces, A is their area (m^2), σ is the Stefan–Boltzmann constant (5.6697×10^{-8} W/m$^2 \cdot$K^4), and $A_1 = A_2$. If the container wall's emissivity (ε_2) is similar to that of the cardboard boxes (Table 4.1) and the surfaces have a line-of-sight view of each other, Eq. (4.8) reduces to:

$$Q_{1-2} = A\sigma(T_1^4 - T_2^4) \qquad (4.9)$$

When the humidity is low enough, evaporative cooling (Section 4.03) may decrease T_1 below T_2 and cause Eqs. (4.8) and (4.9) to yield negative values indicative of the intensity of net outer box to wall radiation (Q_{2-1}). Radiation exchange between the wall and inner boxes is reduced by >90% because inner boxes are shielded from the chamber wall by outer boxes.

Table 4.2 Decline in CO_2 Production by Asparagus, Head Lettuce, Cut Daffodils, and White Rose Potatoes (Hardenburg et al., 1986)

Commmodity	Temperature (°C)	Days in Storage	Air or 1% [O_2]	Respiration (mg CO_2/kg · h)
Asparagus	0	1	Air	60
(Martha Washington)	0	2	Air	46
	0	3	Air	39
	0	4	Air	36
	0	8	Air	32
	10	0	Air	196
	10	1	Air	108
	10	2	Air	83
	10	3	Air	70
Head lettuce	0	1	Air	17
	0	5	Air	9
Iceberg lettuce	10	0.5	Air	25
	10	2.5	Air	16
	10	4.5	Air	15
	10	10	Air	15
Potato (White Rose)	20	2	Air	16
	20	6	Air	11
	20	10	Air	8
Potato (Majestic)	20	0	Air	12
	20	30	Air	7.4
	20	60	Air	5.7
Daffodil	0	1	Air	37.5
	0	2	Air	32.5
	0	3	Air	25.0
	0	8	Air	17.5
	0	20	Air	12.5
Linda carnation	2.2	1	Air	42.0
	2.2	2	Air	38.0
	2.2	4	Air	35.5
	2.2	10	Air	32.0
	2.2	24	Air	26.0
	2.2	1	1% O_2	30.0
	2.2	2	1% O_2	28.5
	2.2	4	1% O_2	24.0

(*Continued*)

Commmodity	Temperature (°C)	Days in Storage	Air or 1% [O₂]	Respiration (mg CO₂/kg · h)
Table 4.2 (Continued)				
	2.2	10	1% O_2	19.0
	2.2	24	1% O_2	15.0

Public domain—in Table 4.2—respiration rate (mg CO_2 produced/kg · h); iceberg lettuce (Platenius, 1942); Majestic potatoes (Burton, 1982); and cut Linda carnations (Uota and Garazsi, 1967) during storage at various temperatures in air or 1% [O_2] at atmospheric pressure. A similar decline occurs in leaves (Blackman, 1953), apples (Kidd and West, 1945), and roses (Siegelman et al., 1958; Coorts et al., 1965).

When a radiation shield, such as Mylar®, is positioned between two radiating surfaces, the heat transfer rate becomes (Özisik, 1985):

$$Q_{1-2} = A\sigma(T_1^4 - T_2^4)/[(1/\varepsilon_1) + (1/\varepsilon_2) - 1) + (1/\varepsilon_{3,1} + 1/\varepsilon_{3,2} - 1)]$$

(4.10)

where $\varepsilon_{3,1}$ and $\varepsilon_{3,2}$ are the emissivities of the Mylar® shield facing the inner surface of the box and plant matter, respectively.

4.3 EVAPORATIVE COOLING

Water loss is a complex phenomenon resulting from mechanical, biological, and physical interactions. Biologists traditionally have formulated the problem in terms of a vapor-pressure gradient between the commodity and air passing over it (Burton, 1982; Ben-Yehoshua, 1986). The larger the vapor-pressure gradient, the more rapidly water is transferred. To diminish water loss, either the relative humidity of the storage air must be increased or the water conductance of the stored commodity decreased, e.g., by waxing the commodity's surface or protecting it with a water-retentive wrap.

Thermodynamics examines water loss in a different, more comprehensive manner. Evaporation can be analyzed as an interaction between three interdependent variables, the availability of latent energy at the evaporating surface, the vapor-pressure gradient which develops at equilibrium, and the resistances in the water–vapor pathway (Raschke, 1960; Slatyer, 1967). When there is a temperature difference in a system, heat flows from regions of high to low temperature, and once the temperature distribution is known the rates of heat transfer can be determined from laws relating heat flux to the temperature gradient. In systems containing biological material, the combined effects of conduction, convection, radiation, and evaporation (or condensation) modulate the vapor-pressure and temperature gradients that develop, and heat transferred by evaporation determines the rate of commodity water loss.

The quantity of water that can be evaporated from an adiabatic system depends on the amount of heat added (first law of thermodynamics). Because water loss results in evaporative cooling, it lowers a stored commodity's temperature unless the latent energy used to change the state of water from liquid to vapor is replaced from a heat source. Therefore, commodity water loss in a refrigerated space depends on respiratory heat, sometimes augmented or reduced by additional heat transferred to or from the stored product by convection or radiation. Respiratory heat is immediately available for this purpose since it is generated within each cell and does not have to be acquired from the environment. When a commodity remains at a constant temperature, if the heat necessary to evaporate transpirational water is less than the total respiratory heat, the commodity is transferring heat to its environment by convection and radiation, and if the heat used to transpire water exceeds the respiratory heat, the commodity is acquiring heat from its environment (Gac, 1956). A commodity stored in a refrigerated space cannot remain at a constant temperature and lose more water than its respiratory heat is capable of vaporizing unless it is colder than the environment and receiving heat from it.

Aerobic respiration generates approximately 9% of the liquid water that must be evaporated to remove all respiratory heat by evaporative cooling (Eq. (4.16)). Additional water vaporized to dispel respiratory heat and any heat acquired from the environment must be drawn from the cellular reserve. During LP, CA, or NA storage, the rate of respiratory heat production progressively declines (Table 4.2), and this causes the rate of water loss to decrease.

The pressure dependence of the resistance to water vapor flux through air-filled stomata, lenticles, pedicle end-scars, barrier air layers, and the intercellular system is given by:

$$r_A = r_{A,R} \frac{\ln[(p_R - p_{v,o})/(p_R - p_{v,i})]}{\ln[(p - p_{v,o})/(p - p_{v,i})]} \qquad (4.11)$$

where p_R is the reference pressure (atmospheric pressure); $r_{A,R}$ the pore resistance (s/cm) at atmospheric pressure for the particular vapour-pressure values $p_{v,i}$ and $p_{v,o}$ inside (i) and outside (o) the surface, respectively; and r_A is the pore resistance at pressure p (Burg and Kosson, 1982, 1983).

Evaporative heat transfer, Q_v (kcal/s) from a commodity to air is given by:

$$\dot{Q}_v = m_v H_v = \left[\frac{A(\Delta p_v H_v)}{r_{v,c,p}}\right]\left[\frac{M_v}{R_u T}\right] \tag{4.12}$$

where m_v (kg) is the weight of water (v) evaporated, H_v (kcal/kg) is the latent heat of water evaporation [595.4 kcal/kg at 0°C; 590.2 kcal/kg at 10°C; 584.9 kcal/kg at 20°C], A (m²) is the commodity's surface area, M is the molecular weight of water ($=18$), T is the temperature (K), $r_{v,c,p}$ is the combined transpirational resistance of the cuticle and pores (s/m), Δp is the water–vapour-pressure gradient between the plant matter and air (atm), and the gas constant R_u equals 0.08295 m³ · atm/kg · mol · K. A horticultural commodity's stomata normally are closed in darkness, but usually they open in LP (Section 3.03), and this decreases the value of $r_{v,c,p}$ in Eq. (4.12).

Cuticular tanspirational resistance is mainly due to soft cuticular waxes that form overlapping hydrophobic platelets surrounding air-filled pores and microcapillaries. The driving force for cuticular transpiration is the water-pressure gradient across the continuous non-porous layer of soluble cuticular lipid (SCL) that separates liquid water and vapor at the interface in cuticular membranes. Evaporation at the outer cuticular surface creates a water potential difference that draws liquid water by mass flow, through cell wall microcapillaries, to epidermal cells at the cuticular surface, and through the liquid water phase of the cuticular membrane, to the air/water interface (Schönherr, 1976; 1982; Schönherr and Schmidt, 1979).

When a commodity is enclosed in a box, heat transfer (Q_v) by water vapor flow through the box walls to the atmosphere is given by:

$$\dot{Q}_v = m_v H_v = \left[\frac{A(\Delta p_v)H_v}{r_{v,box}}\right]\left[\frac{M_v}{R_u T}\right] \tag{4.13}$$

where the box's resistance to water vapor transfer ($r_{v,box}$, s/m) depends on the storage pressure and water−vapor pressure according to the expression:

$$r_{box} = r_{box,R} \frac{\ln[(p_R - p_{v,o})/(p_R - p_{v,i})]}{\ln[(p - p_{v,o})/(p - p_{v,i})]} \tag{4.14}$$

and $r_{box,R}$ is the box resistance measured at reference pressure P_R (atm) for vapour-pressure values $p_{V,i}$ and $p_{V,o}$ inside (i) and outside (o) the box, and p (atm) is the storage pressure.

Hypobaric research ceased in the West nearly three decades ago when postharvest physiologists became convinced that LP causes an excessive and unavoidable commodity water loss by drastically increasing the rate at which water vapor diffuses from plant tissues and boxes (Figure 2.2). A thermodynamic analysis predicts that hypobaric storage should cause a very low water loss if the commodity is prevented from receiving heat from the environment. Roses experienced a 6.78% weight loss during a 35-day storage in a VivaFresh hypobaric warehouse when they ostensibly were the warmest objects in the storage and should have lost at most 2.5% of their weight transferring the small amount of respiratory heat they produced (Table 10.1, upper). Was this discrepancy caused by unusually rapid water vapor diffusion from the roses due to the low pressure (Figure 2.2), as academics proposed? A thermodynamic analysis was performed to determine the cause of the extra weight loss. The flowers lost 0.24% of their weight vacuum cooling from 4.1 to 2.5°C during the initial warehouse pumpdown (Figures 10.2 and 10.3). Subsequently, the nonwaxed cardboard boxes unexpectedly remained warmer than the flowers for 12.5 days, and then this trend reversed. By the 35th day, water condensation in each cardboard box's micropores had increased the cardboard's weight by 18% (Eq. (10.1)). During the initial 12.5 days, heat released from water vapor that condensed in the cardboard was transferred by radiation and convection from the warmer boxes to the cooler flowers, and the flowers dispelled this heat by evaporative cooling. This analysis indicates that the extra commodity water loss was due to heat generated by water condensation in the cardboard, and that respiratory heat only caused an average weight loss of 2.5% in 35 days (see Chapter 10). The calculation was verified by an experiment in which rose storage was compared in equally sized cardboard and plastic boxes. After the initial vacuum cool-down, weight loss from the roses was reduced by 73% in plastic boxes because they lacked micropores that condense water. A Mylar® radiation reflecting liner (Table 4.1) situated between a cardboard box's inner surface and the stored roses caused a large reduction in flower weight loss by interfering with radiation of heat from the cardboard to the roses (Table 10.1, lower).

4.4 CONDUCTION

Conductive heat flow in the x direction is given by the Fourier law:

$$Q_x = -kA(\mathrm{d}T/dx) \qquad (4.15)$$

where Q_x is the rate of heat flow (W) through area A (m^2) in a positive x direction, $\mathrm{d}T/\mathrm{d}x$ is the temperature gradient in the x direction, and the proportionality constant k (W/m · °C) is the material's thermal conductivity. When commodities are tightly packed, metabolic heat causes the temperature to be higher in the center of a box, and Q_x in an outward direction through restricted areas where individual commodity samples contact each other is limited by the small contact area (A) and high "thermal contact resistance" in the contact interfaces between the commodity surfaces. The magnitude of this resistance to conductive heat transfer depends on the interface pressure forcing the surfaces close together and their roughness. Direct contact between the solid surfaces takes place at a limited number of interfaces, decreasing A. During hypobaric storage, heat is primarily transferred by conduction through the vacuum-filled interface voids since convection is markedly inhibited in a vacuum and cannot occur in such a thin layer. As the thermal conductivity of a vacuum is far lower than that of solid surfaces, the interface acts a resistance to heat transfer. The thermal contact resistance cannot be reliably computed and can only be determined experimentally.

4.5 HEAT FORMATION AND ATP PRODUCTION

A mole of ATP contains 7.3 kcal of energy. Fermentation and aerobic respiration are 37.2% and 34.7% efficient in forming ATP, respectively. The rest of the Gibbs free energy lost during aerobic respiration and fermentation is released as heat (Eqs. (4.16) and (4.17)). The latent heats of vaporization are 6.15, 9.85 and 7.71 kcal/mole for acetaldehyde, ethanol, and water, respectively.

Aerobic respiration:

$$\text{glucose} + 32\,\text{ADP} + 32\,\text{P}_i + 6\,\text{O}_{2\,\text{gas}} = 6\,\text{CO}_{2\,\text{gas}} \qquad (4.16)$$
$$+ 6\text{H}_2\,\text{O}_{\text{liquid}} + 32\,\text{ATP} + 439.4\,\text{kcal(heat)}$$

Fermentation:

$$\text{glucose} + 2\,\text{ADP} + 2\,\text{P}_i + 2\,\text{NAD} = 2\,\text{CO}_{2\,\text{gas}} + 2\,\text{CH}_3\text{CHO}_{\text{liquid}}$$
$$+ 2\,\text{NADH} + 2\,\text{ATP} + 33.5\,\text{kcal(heat)}$$
$$2\,\text{CH}_3\text{CHO}_{\text{liquid}} + 2\,\text{NADH} = 2\,\text{CH}_3\text{CH}_2\text{OH}_{\text{liquid}}$$
$$+ 2\,\text{NAD} + 5.7\,\text{kcal(heat)}$$

$$(4.17)$$

4.6 REMOVING RESPIRATORY HEAT FROM A HYPOBARIC CHAMBER

After cool-down has been completed, a commodity in a hypobaric apparatus can only remain at a constant temperature if its respiratory heat equals the net heat loss (Section 4.3). Tests performed by Grumman in a Dormavac hypobaric intermodal container and in small bell jars indicated that at a hypobaric pressure the low film coefficient for natural convection and restricted heat-carrying capacity of a low-density air/water—vapor mixture make it difficult, regardless of the available refrigeration capacity, for an LP heat exchange system to transfer a densely packed horticultural commodity's respiratory heat to the refrigeration source as rapidly as it is produced:

- The heat-source/heat-sink surface area ratio is nearly unity in a laboratory hypobaric chamber, and this results in rapid cooling by radiant-heat transfer from exterior commodity to a laboratory vessel's walls. Exterior boxes in a fully loaded intermodal container are cooled in part by radiation, but interior boxes are not since they lack a view of the container wall.
- Evaporative cooling due to commodity moisture loss is the predominant heat transfer mode in an LP intermodal container.
- Convective cooling is ineffective at a low pressure because the air's low density reduces its specific heat and the convective heat transfer coefficient.

Example: After papayas with more than a 50% at-harvest color break were heat-treated for insect quarantine and re-cooled to 10°C, they could not be shipped in Dormavac hypobaric intermodal containers packed in cartons sealed for insect control. The extra heat produced during their respiratory climacteric caused the fruit's temperature to increase to 22°C during transit. At 22°C the fruit's vapor pressure equaled the 20 mmHg storage pressure and vacuum cooling prevented a further temperature rise. Fermentative low-O_2 damage resulted because the papayas' interior had become anaerobic. In the same hypobaric shipment heat-treated papayas with a color break of less than 50%, packed in similar sealed boxes, remained close to the 10°C storage temperature, and the fruits were well preserved. This example illustrates the importance of ensuring an adequate respiratory heat transfer capability under hypobaric conditions.

Jiao et al. (2012) recommended that the exterior surface of laboratory hypobaric storage chambers located in a cold storage room within which the air temperature fluctuated by ±1.1°C every 40 min, should be covered with 13 mm (0.51 in) thick foam insulation to stabilize the LP chamber's interior air temperature during insect disinfestation studies. This thickness of insulation reduced the wall temperature variation in an empty chamber to ±0.2°C. Jiao et al. did not provide measurements of the chamber's temperature variability without insulation, but if the room's air temperature changed by 1.1°C at a linear rate during each 20 min ON or OFF portion of the refrigeration cycle, the chamber wall temperature would have fluctuated by less than ±0.79°C. In a cold room in which 13°C refrigerated air cycled by ±0.5°C, an identical LP chamber's wall temperature only varied by ±0.2°C without insulation. Radiation from the commodity to the chamber wall is the major heat transfer mode removing respiratory heat in laboratory hypobaric chambers. Insulating a chamber to prevent inward heat transfer also inhibits outward heat transfer from the chamber wall to the refrigerated air in the cold room or enclosure within which the laboratory chamber is located. As the incoming air change in laboratory systems is mechanically saturated at the storage temperature, the transpiration rate needed to remove the respiratory heat can only occur after the commodity's temperature has increased sufficiently to provide the vapor-pressure gradient needed to evaporate commodity water at the required rate. The chamber air warms and transpired water vapor supersaturates the storage atmosphere when the commodity's temperature increases. This makes it difficult to establish the required vapor-pressure gradient. If the pressure is controlled with a vacuum breaker (Section 5.03; Figure 5.2, left) transpired moisture is not removed and instead condenses on the chamber wall, transferring the respiratory heat into the wall, increasing the walls temperature without removing the heat, In Jiao et al.'s LP system, the pressure was controlled by a heated vacuum regulator (Section 5.03; Figure 5.1, right) and the pumping speed should accelerate to remove the transpired water, decreasing the chamber air's pO_2 below the value computed by Jiao et al. (see Chapter 6). To ensure adequate heat transfer a laboratory hypobaric chamber housed in a cold room should not be insulated. Instead, if it is necessary to reduce the cold room's temperature variability, the room's refrigeration system should be modulated.

Chambers similar to those used by Jiao et al. have been adapted for installation in an air conditioned 20–25°C laboratory room by bonding rectangular aluminum cooling channels of appropriate hydraulic diameter at approximately 1 ft intervals to each chamber's walls, door, and humidifier using a heat conducting aluminum bonding agent. All surfaces and piping are covered with 4 in of closed-cell insulation, and each chamber is cooled by a modulated glycol chiller with sufficient capacity to handle the respiratory heat load and all infiltrating heat when the chamber is fully loaded with plant matter. In this system, the chamber walls vary by only ±0.1°C when the temperature is controlled between 0°C and 16°C.

CHAPTER 5

Materials and Methods

5.1 MEASURING THE RH

The most accurate commercially available humidity measuring devices have a precision of $\pm 1.7\%$ RH in the 95–99% RH range. This is not sufficient for LP monitoring or control because the error spans the entire range between a small water loss and extreme desiccation during hypobaric storage. Therefore, in laboratory studies and hypobaric intermodal containers and warehouses, a wet- and dry-bulb temperature measurement with $\pm 0.05°C$ thermistors shielded from radiation with Mylar® (Table 4.1) has been used to compute the chamber RH with an accuracy of $\pm 0.1\%$ at 99.9% RH. The wet-bulb measurement corresponds to the chamber dew-point temperature since heat transfer to the wet-bulb by convection is limited by the low chamber pressure (Section 4.1), and Mylar® limits heat transfer by radiation (Section 4.2). The RH is computed from the dew-point and dry-bulb temperature measurements. Accuracy decreases at higher pressures because heat transfer by convection from the chamber air to the wet-bulb increases.

5.2 MEASURING THE PRESSURE

Pressure is measured with a temperature compensated absolute capacitance transducer. These devices vary in accuracy between approximately ± 0.1 and ± 0.6 mmHg depending on the transducer's span.

5.3 CONTROLLING THE PRESSURE

A vacuum *breaker* controls the pressure in a chamber by adjusting the rate at which atmospheric air enters while the vacuum pump withdraws low-pressure air at a constant rate. A vacuum *regulator* controls the pressure by adjusting the rate at which the vacuum pump evacuates the chamber while atmospheric air enters at a constant rate. Hypobaric intermodal containers and warehouses have used vacuum breakers;

Hypobaric Storage in Food Industry. DOI: http://dx.doi.org/10.1016/B978-0-12-419962-0.00005-X

laboratory LP systems have been equipped with either vacuum regulators or breakers (Jamieson, 1980a,b; Kader, 1975; Jiao et al., 2012; Davenport et al., 2006; Burg, 2004). Static systems also have been employed, in which pumping ceases after the desired pressure has been reached and intermittently resumes to return the chamber to the set pressure after air has been intentionally reintroduced or has leaked into the chamber (Zhang et al., 2005a,b; Li et al., 2004; Li and Zhang, 2005; Schuerger and Nicholson, 2006; Schuerger et al., 2013).

Spring-loaded vacuum breakers control the pressure at their outlet port (Figure 5.1, left). The adjustment knob sets the spring pressure on the device's diaphragm, usually with atmospheric pressure on the upper side and process pressure registering below the diaphragm. The spring pulls the diaphragm up, closing the valve by forcing the plunger against the valve seat. During pump-down, the breaker is closed until the process pressure decreases below the vacuum set-point established by the spring pressure, and then the plunger forces the valve seat downward, allowing outside air to enter. This causes the process pressure to increase toward the set-point, until the plunger rises again to restrict the inflow of air.

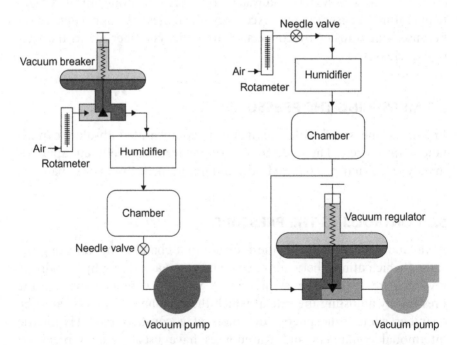

Figure 5.1 Vacuum control systems used for hypobaric storage: (left) vacuum breaker; (right) vacuum regulator.

Vacuum regulators maintain a constant process pressure at their inlet by throttling flow from their outlet to the vacuum pump. Their diaphragm is spring-loaded with atmospheric pressure on the upper side and process pressure on the lower side. The spring opens the valve by pushing the diaphragm downward. The adjustment knob varies the spring pressure to set the process vacuum level below which the valve plug opens and allows the vacuum pump to return the vacuum to the set pressure. During pump-down, the regulator remains open until the set pressure is reached.

Vacuum regulators and breakers may be "self-relieving," providing a small amount of in-leakage to elevate the pressure if it decreases below the vacuum set-point. This is unnecessary when vacuum breakers are used for hypobaric storage since the process provides controlled in-leaking air (Figure 5.2).

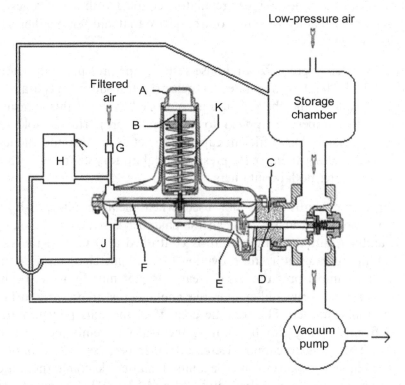

Figure 5.2 Absolute vacuum regulator computer-controlled by a proportional solenoid control valve responsive to an absolute pressure transducer. A, closing cap; B, spring tension adjusting nut; C, blocked throat; D, O-ring seal on valve stem; E, lever assembly; F, diaphragm; G, orifice to make proportional solenoid valve (H) self-relieving (not required with a manual self-relieving vacuum controller); H, proportional solenoid valve (can be computer controlled) or a self-relieving vacuum breaker (manually controlled); J, external register; K, spring.

Hypobaric intermodal containers operate intermittently over-the-road at altitudes that cause a vacuum breaker to close or a vacuum regulator to open if the regulator or breaker has a barometric reference pressure. To prevent this from occurring, the reference pressure has in the past been controlled ± 0.3 mmHg by a low capacity self-relieving nullamatic *absolute pressure* vacuum regulator, which has a sealed absolute pressure reference. A blocked throat and O-ring stem seal isolated the reference side of a large capacity spring-loaded vacuum breaker's diaphragm, and an external registration was connected from the breaker to the storage chamber to sense and control the pressure. A 100 mmHg bias pressure permanently applied to the main breaker by its adjustable spring allowed the regulator's adjustment knob to set the process pressure to values lower than 10 mmHg, even though by itself it could not control pressure below 20 mmHg. This system has been replaced in LP laboratory systems by a regulator that is a combination of an absolute pressure regulator, coupled with a self-relieving built-in large capacity vacuum control device that can serve either as a breaker or regulator.

In Vivafresh hypobaric warehouse systems, the pressure is measured with a 1–1000 mbar absolute piezoelectric transducer, accurate to <1% of reading, and the pressure is controlled $\pm 1\%$ of this measurement by the amperage applied to a compact proportional solenoid valve. This system has sufficient capacity to act as an absolute vacuum breaker controlling the process pressure in laboratory chambers, warehouses, and intermodal containers.

A pressure controller is not used when meat, poultry, fish, or shrimp is preserved in LP. Instead the chamber is sealed and isolated from ambient air, the temperature is controlled at $0°C$, and the vacuum pump operates at a rate providing 1–2 volume changes per hour. When the chamber pressure has decreased to 4.58 mmHg, water begins to boil from the tissue and evaporative cooling crust freezes the surface of the animal matter. The osmotic content of the cells prevents their liquid from freezing and perforating the plasma membrane with ice crystals. The pressure cannot decrease further because water continuously sublimes or boils from the animal matter, keeping the water vapor pressure in the chamber saturated at 4.58 mmHg. An anaerobic condition results in which metabolic odors produced by the commodity and any in-leaking $[O_2]$ are continuously flushed away by cold steam.

Depending on the pumping speed, the animal matter's weight decreases by less than 1–2% per month in a fully loaded chamber.

5.4 VACUUM PUMP

A single-stage, oil-sealed, rotary veined, air-cooled vacuum pump is used in all hypobaric systems. Before the LP mixture enters the vacuum pump, the 0–15°C process gas's temperature increases to 20–25°C, and the low-pressure air's RH decreases by 53–81%. Sufficient "gas ballast" air is admitted to ensure opening of the vacuum pump's exhaust valve before the water condensation pressure is reached. With full gas ballast, a single stage vacuum pump typically is >90% efficient at a 4.6 mmHg suction pressure.

5.5 MEASURING THE AIR-CHANGE RATE

The air-change rate is determined by the pumping speed of the vacuum pump and can be measured at atmospheric pressure by means of a flow meter located upstream of a vacuum *breaker*. When an absolute vacuum breaker is controlling the process pressure, the air-change rate can be reduced by throttling the vacuum pump at its suction inlet with a flow control means or by controlling the pump's rpm.

Air-flow through a system employing an absolute vacuum *regulator*, operating at 20–25°C, cannot be calculated based on the rate at which atmospheric air enters a mechanically humidified hypobaric system because atmospheric air continuously enters the system at the initial rate, but the regulator increases (or decreases) the flow rate from the chamber to the vacuum pump in response to transpired water and the altered air and water vapor pressures which develop at the elevated regulator temperature (see Chapter 6). The flow rate from the chamber to the vacuum pump can be measured by temporarily operating the vacuum pump without gas ballast and determining the time required to collect a fixed quantity of air piped from the vacuum pump's exhaust into an inverted water-filled graduated cylinder. The water vapor partial pressure in 20–25°C air entering a laboratory LP system at atmospheric pressure is relatively small, and the water partial pressure is equally low in air collected by slowly bubbling the pump exhaust through 20–25°C water. Therefore, the water partial pressure in the incoming and outgoing air does not significantly influence the result, and the rate of flow from the chamber to the pump can be computed by comparing the measured rate at which air enters and leaves the system.

5.6 MEASURING O_2, CO_2, C_2H_5OH, AND CH_3CHO IN THE AIR CHANGE

Total CO_2 production and O_2 consumption due to respiration and fermentation is measured during hypobaric storage by comparing the $[CO_2]$ and $[O_2]$ concentrations present in air entering the hypobaric chamber vs. their concentrations in air exhausted from the vacuum pump. The vacuum pump is temporarily operated without gas ballast, and because O_2 and CO_2 have a very low solubility in vacuum pump oil, they readily pass through into the pump's exhaust where the $[CO_2]$ increase and $[O_2]$ decrease is amplified in proportion to the pump's compression ratio. This creates CO_2 and O_2 concentrations in the pump's exhaust which can be easily measured by gas chromatography or with colorimetric gas detector tubes (Section 5.9).

The rate at which ethanol and acetaldehyde vapors emanate from the commodity cannot be measured after these vapors pass through the vacuum pump because they readily dissolve in vacuum pump oil. To determine their concentration in the air change, a bypass pipe containing a 500 mL vacuum-tight chamber was connected in parallel to the main vacuum pump suction line. After this sampling chamber's volume equilibrated with the flowing air change, inlet and outlet valves to the chamber were closed and the chamber was vented with bone-dry CO_2-free air, its contents were thoroughly mixed by means of an integral miniature "computer" fan, and samples were removed with a gas-tight syringe for analysis by gas chromatography. The GC measurement is the VOC's concentration in the sample chamber at the storage pressure and air-change rate (Table 5.1).

Table 5.1 Henry Law Coefficients for Ethanol, Acetaldehyde, and Carbon Dioxide at 373.15 K (100°C)			
	$K_H°$ (mol/ kg/bar)	$\exp[d(\ln K_H)/d(1/T)]$	$K_H^{(373.15\,K)}$ (mol/kg/bar)
Ethanol	190	6600	2.888
Acetaldehyde	15	5700	0.234
Carbon dioxide	0.034	2400	0.0062
$K_H°$ and $[d(\ln K_H)/d(1/T)]$ are from Sander (1999), and $K_H^{373.15\,K}$ is computed from Eq. (5.1).			

5.7 MEASURING O_2, CO_2, C_2H_5OH, AND CH_3CHO WITHIN THE COMMODITY

Full evaporation head-space gas chromatography (HSGC) was used for these determinations. To prevent changes from occurring during the measurement, tissue cylinders were cut from the commodity immediately after an LP chamber was vented, they were trimmed to a specific length, inserted into preweighed 20 mL headspace sampling tubes, preweighed caps were crimped on, the tubes were reweighed to determine the tissue weight, and the tubes were immediately autoclaved for 20 min at 15 pounds pressure to sterilize the tissue and inactivate its enzymes. The tubes were then placed into a beaker filled with water and transferred into a microwave oven. The oven was powered until water in the beaker had boiled for at least 1 min, warming the tissue to 100°C. At that temperature $98.1 \pm 1.76\%$ of the tissue ethanol is present in the headspace tube's air phase (Zhu and Chai, 2005; Li et al., 2009). The Henry's law coefficients (K_H, mol/kg/bar) for acetaldehyde and carbon dioxide at 372.15 K (100°C) can be calculated from the expression:

$$K_H^{(T)} = K_H°\exp[d(\ln K_H)/d(1/T)](1/T - 1/298.15) \qquad (5.1)$$

where $K_H°$ is the Henry's law constant at 298.15 K (25°C). Nearly 100% of the tissue's acetaldehyde and carbon dioxide must have partitioned into the HS tube's air phase since at 100°C the $K_H^{(373.15\ K)}$ constant for these substances is lower than that for ethanol. When the commodity's pH is 5 or lower, dissolved CO_2 is measured by this assay without contamination from cellular carbonate or bicarbonate (Table 5.2).

After a headspace tube is removed from the microwave oven, a gas sample is immediately withdrawn from it with a 1 mL gas-tight syringe

Table 5.2 Influence of pH on the Partitioning of CO_2, HCO_3^-, and H_2CO_3 Between an Aqueous Solution and an Adjacent Air Phase at 20°C and Atmospheric Pressure	
pH	$[CO_2 + HCO_3^- + H_2CO_3]/[CO_2]$ in Aqueous Phase
4	0.91
5	0.96
6	1.48
7	6.6
8	58.0
Data from Stumm and Morgan (1981).	

and injected into the gas chromatograph. At 100°C, the pressure in each HS tube prepared from 13°C tissue increases to 2.2 atm due to air expansion and water saturation. Therefore, 54.5% of the sample is vented and lost to the atmosphere when the syringe needle is removed from the HS tube, before the sample is injected into the GC.

5.8 FLOW CONTROL

The rate at which air enters a VacuFresh hypobaric intermodal container is set at a fixed value by a flow controller when a vacuum regulator is used to control the pressure. Because the downstream/upstream pressure ratio is <0.53, flow through the controller is "choked" (sonic) and depends on the upstream pressure. The vacuum regulator has to continuously vary the pumping speed to compensate for natural shifts in barometric pressure and decreases in barometric pressure due to elevation when the container is operating over-the-road or by rail.

In a redesigned version of VacuFresh, in which a vacuum regulator controls the pressure (Figures 13.2 and 13.3), a thermal mass flow controller prevents flow fluctuations caused by shifts in barometric pressure. Metered air is divided into two laminar flow paths in the thermal mass flow controller, one through the primary flow conduit, the other through a capillary sensor tube. Both conduits are designed to ensure laminar flow and therefore the ratio of their flow rates is constant. Two precision temperature-sensing windings on the sensor tube are heated, and when flow takes place, gas carries heat from the upstream to the downstream windings. The resultant temperature differential is proportional to the change in the resistance of the sensor windings. A Wheatstone bridge design monitors the temperature-dependent resistance gradient on the sensor windings, which is linearly proportional to the instantaneous rate of flow. Output signals of 0−5 Vdc or 4−10 mA are generated, indicating the molecular-based flow rates of the metered air. The combined gas streams flow through a proportionating electromagnetic valve with an appropriate orifice. The mass flow rate is unaffected by temperature and pressure within stated limits.

5.9 MEASURING HYPOBARIC ACID VAPOR

The concentration of hypochlorous acid vapor (HOCl) present in a chamber or emitting from the hypochlorous acid generating system

(Section 12.7) was measured at atmospheric pressure using a Kitigawa air sampling pump that draws 50 or 100 mL of air per pump stroke through a #109SB chlorine detector tube (0.1–10 ppm). The HOCl concentration in an evacuated (low pressure) chamber was determined by withdrawing an air sample after the chamber was vented to atmospheric pressure. The #109SB chlorine detector tube also measures chlorine and chlorine dioxide gases, but these do not interfere with the HOCl measurement because they are not generated by the HOCl emission system (Section 12.7).

The same gas sampling system can also be used to measure O_2, CO_2, ethanol, and acetaldehyde in the air change using appropriate gas sampling tubes (Section 5.6).

CHAPTER 6

Humidity Control

During hypobaric storage a "metabolic humidification system" evapo-rates sufficient cellular water to transfer most of the respiratory and fermentative heat and additional heat that plant matter acquires from the environment (Section 4.3). In LP warehouses and laboratory systems, the metabolic humidifier is supplemented with a mechanical system utilizing electric heat to evaporate sufficient ancillary water to saturate the incoming air-changes. The advantage gained with a mechanical humidifier is that a full commodity load, partial load, even a single fruit can be stored, certain that the humidity will always be saturated to minimize commodity water loss.

Attempts have been made to control the relative humidity (RH) at a high value in Western (Tolle, 1969, 1972) and Chinese (Li et al., 2004; Li and Zhang, 2005) laboratory chambers responsive to a humidity sensor, and a hypobaric warehouse design envisioned a simi-lar arrangement (Anon, 1974). Tolle (1969, 1972) stored strawberries and tomatoes in an LP apparatus without humidification and found that "drying of the fruits increased at faster airflow rates even though electro-sensors recorded the same high RH irrespective of the airflow rate." Dilley and Dewey (1973) preserved apples with negligible weight loss for nearly 1 year in humidified LP chambers, but Lougheed et al. (1977) reported that apples desiccated even though a humidity sensor indicated that the rarified air in the chamber was nearly saturated. These anomalies were clarified by proprietary tests carried out in a Grumman/Dormavac hypobaric intermodal container in which the humidity was controlled by a bureau of standards chilled-mirror dew-point sensor. Whenever the humidity decreased below 95% RH, a low-pressure water boiler's electric immersion heater energized to inject cold steam into the incoming low-pressure air change. Without cargo present the system worked as envisioned, but after the container was filled with 30,000 pounds of plant matter the humidification heater failed to energize. Metabolic heat evaporated commodity water so rap-idly through the stored plant matter's enormous surface area that evaporative cooling reduced the product's temperature and caused

Hypobaric Storage in Food Industry. DOI: http://dx.doi.org/10.1016/B978-0-12-419962-0.00006-1

heat to transfer by radiation and convection from the warmer container walls to the cooler commodity. This heat evaporated enough additional commodity water (Table 2.1) to keep the chamber RH above 95% and prevent the boiler's water heater from energizing. To prevent an excessive commodity water loss, the RH sensor was abandoned and the boiler heater was continuously energized at a constant wattage high enough to saturate the incoming low-pressure air change.

A chilled-mirror dew-point sensor was used in the Grumman test at its highest reliable noncondensing upper limit of 95% RH, where its accuracy was ±1.25% RH. Because water vapor diffusion from plant matter is accelerated at a low pressure (Figure 2.2), the RH must be controlled at a higher value than 95%, close to 99.5−99.8% (Section 2.1) to minimize commodity water loss in LP. Therefore, Grumman's test does not preclude controlling humidity using wet- and dry bulb ±0.05°C thermistors shielded from radiation with Mylar®, to control the humidity at a high enough value to minimize commodity water loss. The thermistor system has an accuracy of ±0.1% RH at 99.8% RH (Section 5.1). The high RH target can be computer controlled by a method similar to that described in Burg et al. (2009). A vacuum regulator (Figure 5.2) prevents condensation (see Chapter 7) and continuously controls the pressure by regulating the pumping speed (Section 5.2), while the air-change rate is continuously adjusted with a thermal mass flow controller (Figure 13.1-#32 and Figure 13.2-M) to produce the target RH from transpired commodity moisture. For optimal performance and reliability at a high humidity and severe environmental exposure, the thermistors should be a glass-encapsulated ±0.05°C negative temperature coefficient (NTC) type. The thermistor wiring penetrates the low-pressure zone through pin-connector-type hermetic vacuum seals that prevent leakage.

A mechanical humidifier was not used in VacuFresh because intermodal containers only ship full loads, and therefore can be designed to use metabolic humidification. Relying solely on metabolic humidification has the advantage that it eliminates the necessity to carry water, lowers energy consumption, and reduces cost (see Chapter 13). In the original VacuFresh design, a vacuum breaker (Section 5.3; Figure 5.1, left) controlled the pressure by regulating the rate at which atmospheric air entered the vacuum tank through a pneumatic air horn's collar (Figure 6.1; Figure 13.2-J; Burg, 1987a,b). The incoming air

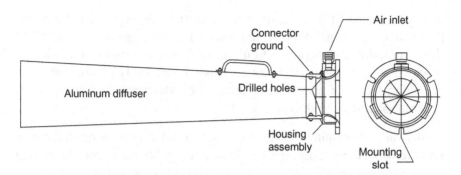

Figure 6.1 Pneumatic air mover. The jets have been modified to provide optimal flow in a 20 ft VacuFresh hypo-baric intermodal container. STM permission to use Figure 11.11—Pneumatic air horn, in Burg (2004), granted to Elsevier.

pressurizes the air horn's jets, inducing circulation of up to 40 volumes of low-pressure chamber air per volume of expanded incoming dry air. Discharge from the air horn flows though a duct to the opposite end of the container (Figure 13.2-K) and returns to the air horn's suction after passing through cargo boxes stagger-stacked to facilitate longitudinal airflow (Figure 13.3). The return air, saturated by transpired commodity water, mixes with expanded dry incoming air in the pneumatic air mover and elevates the RH to approximately 97% in the discharge. Without producing mechanical heat, the pressure difference created by the vacuum pump powers the air horn, causing it to emit up to 2500 cfm of low-pressure cargo air at sonic velocity. Simultaneously, the vacuum pump induces downward laminar airflow into a floor-level vacuum pump suction duct configured with holes sized and longitudinally spaced to induce equal uptake of rarified chamber air along the duct's entire length (Burg and Kosson, 1983). To minimize commodity water loss, the vacuum pump is manually throttled to cause product water evaporated in response to metabolic heat produced by the type and weight of plant matter being shipped, to saturate the incoming air change after expansion dried the air during entry into the storage chamber (Burg, 1987a,b). The amount of commodity water escaping from a VacuFresh container depends on the pumping speed and the chamber air's moisture content at the storage temperature (Table 9.1). The appropriate pumping speed for different types of plant matter varies between 0.9 and 3.2 air changes per hour, based on the amount of metabolic heat the particular commodity produces at the storage temperature and pressure (Burg, 2004). Metabolic [CO_2] and [C_2H_4] are prevented from accumulating, and respiratory [O_2] drawdown does not significantly lower the chamber's [O_2] partial pressure because the selected air-change rate depends on the commodity's CO_2 production and O_2 consumption rates. Lack of a mechanical humidifier does not increase the quantity of moisture that is evaporated from plant matter since that is determined by the amount of metabolic heat produced.

Because the original VacuFresh intermodal container operated at a predetermined pumping speed, the reduction in transpiration caused by a progressive decline in metabolic heat production (Table 4.2) reduced the chamber dew point and caused the commodity's temperature to decline (Section 4.3). Radiation and convection then transferred heat from the warmer chamber atmosphere and walls to the cooler commodity, stimulating extra transpiration, which returned the chamber RH to near-saturation. In spite of a reduced rate of metabolic heat production the commodity still lost water at the initial rate determined by the pumping speed and water content of saturated chamber air at the storage temperature.

CHAPTER 7

Water Condensation in Hypobaric Chambers

It has been assumed that vacuum breakers and regulators are equally effective controlling the pressure in LP systems (Figure 5.1; Section 5.3). Vacuum regulators have been employed in only a few laboratory systems (Jamieson, 1980a,b; Kader, 1975; Davenport et al., 2006; Burg, 2004; Jiao et al., 2012); vacuum breakers were used in Dormavac (Figure 1.1), Fruehauf, and VacuFresh (Figure 1.2) hypobaric intermodal containers, Grumman/Dormavac hypobaric warehouses (Burg, 2004), Vivafresh warehouses (Figure 1.3), and in most laboratory LP systems.

Water transpired by plant matter supersaturates a hypobaric storage chamber if the incoming air change has been saturated at the storage temperature by a mechanical humidifier (Chapter 6). The system's pressure control device, pressure measuring transducer, and vacuum pump are operated at 20–25°C to ensure that process water in air saturated at the storage temperature does not condense in these instruments and interfere with their operation. The high buoyancy of the chamber's supersaturated low-pressure mixture causes it to rise (Eqs. (4.3) and (4.4); Burg and Kosson, 1982, 1983), and when a vacuum breaker is used to control the pressure (Section 5.3; Figures 5.2 and 6.1, left), the excess water vapor condenses on the storage chamber's roof and walls. In mechanically saturated systems controlled by vacuum breakers, transpired product moisture condensed under the lids of glass vacuum desiccators (Burg, 2004), on transparent viewing plates in steel laboratory vacuum chambers (Lougheed et al., 1978), in Atlas Technology leak-tight aluminum vacuum chambers, and under the roof of Fruehauf and Dormavac hypobaric intermodal containers and Vivafresh warehouses.

Transpired water did not condense and accumulate in laboratory hypobaric chambers during 8-week full-load tests with various types of plant matter when a mechanical humidifier saturated the incoming low-pressure air change and the pressure was controlled with a vacuum

Hypobaric Storage in Food Industry. DOI: http://dx.doi.org/10.1016/B978-0-12-419962-0.00007-3

regulator operating at 25°C. During full-load tests in the same laboratory chamber at an identical storage temperature, pressure, and mechanically humidified air-change rate, a substantial quantity of transpired water condensed when the pressure was controlled with a vacuum breaker operating at 25°C. This unexpected difference between hypobaric storage systems operating with vacuum breakers and regulators has not been reported in prior art and is described for the first time in US provisional patent application No. 6170501 (Burg, 2012).

If low-pressure air supersaturated by a combination of mechanical humidification and commodity transpiration escaped before condensing in a hypobaric storage chamber, the rarified air's low heat capacity would cause it to rapidly warm and the pressure to increase upstream of a vacuum regulator operating at 25°C. The regulator would respond by enhancing the pumping speed to offset the pressure rise, while water vapor continued entering the storage chamber at the initial rate determined by the humidification system's wattage setting, and air entered at the initial rate determined by choked flow through a preadjusted flow-regulating restriction. The system would equilibrate when the increased rate of air and moisture flow from the chamber to the vacuum regulator and vacuum pump decreased the chamber RH to saturation, preventing condensation and stabilized the pressure in the storage chamber. This explanation was confirmed by measurements demonstrating that when one saturated 15 mmHg, 13°C air change per hour entered a 170-L hypobaric chamber containing 31.3 kg of mangos, a vacuum regulator operating at 25°C increased the rate at which the low-pressure mixture flowed from the chamber to the vacuum pump by 1.54-fold (Table 5.2).

Vacuum breakers control the rate at which air enters a storage chamber. If air supersaturated by commodity transpiration and mechanical humidification escaped from an LP storage chamber in which the pressure was controlled with a vacuum breaker, and this mixture warmed to 25°C, elevating the pressure upstream of the vacuum pump, the breaker would sense the pressure rise which back-fed into the storage chamber and try to decrease the pressure by slowing the rate at which air entered the storage chamber. The initial amount of water vapor would continue entering accompanied by less air, and the excess moisture would condense in the chamber.

It is uncertain how supersaturated water vapor escapes from an LP storage chamber controlled with a vacuum regulator. Mist or fog might form in the storage chamber and vaporize after being carried over by the air change into the 25°C suction pipe leading to the vacuum regulator and pump, or supersaturated air might flow directly from the storage chamber to the vacuum pump.

At atmospheric pressure, fruits transpire water at a significant rate into saturated ambient air (Lentz and Rooke, 1964; Wardlaw and Leonard, 1940). Respiratory heat warms the fruit above the air temperature, creating a fruit to air vapor pressure gradient that evaporates (transpires) water vapor into the saturated air. The evaporated water vapor is cooled and condensed by convection, causing mist to form (Rashke, 1960). Mist can arise in air either by heterogeneous or homogeneous condensation. The mist formation described by Rashke is the heterogeneous type that occurs on natural or man-made cloud condensation nuclei (CCNs) such as volcanic dust, clay, soot, black carbon, sea salt, ammonium sulfate, and certain organic aerosols. Normally there are 100–1000 CCN per cubic centimeter of atmospheric air, but in a hypobaric storage chamber the CCN frequency is decreased proportionate to the reduction in air partial pressure. Only 1.3–26.3 CCNs per cubic centimeter are present at a 10–20 mmHg storage pressure. This slows or prevents heterogeneous condensation.

Homogeneous condensation occurs in air lacking CCNs, where droplet formation and growth depend on statistical collisions between water molecules. Approximately 400–600% supersaturation is needed for homogenous condensation of pure water vapor to begin because newly formed water droplets are extremely small. Therefore, homogeneous condensation does not occur during LP storage.

Water vapor also may condense and form fog or mist adjacent to a heat exchanger's surface if it is at least 2.5°C colder than the adjacent water-saturated air layer and the air in the stagnant surface layer cools faster than water can be removed from it by mass transfer (Brouwers, 1990, 1992a,b; Brouwers and Chesters, 1992; Saleh, 2004). The low air velocity, typical of LP storage, is one of the most important factors causing mist or fog to form on a heat exchanger's surface at atmospheric pressure (Koyama et al., 2002). Brouwers (1990, 1992a,b) and Brouwers and Chesters (1992) used tangency condition models to study fog formation in cooler condensers at atmospheric pressure. They determined the vapor mole fraction and temperature in the stagnant wall film and compared this value to the water saturation line in the air/water vapor mixture to determine possible crossing of this line. The part of the film positioned in the supersaturated region is thermodynamically unstable and condensation is likely to occur at vapor mass fractions between 0.05 and 0.45 or higher, creating fog in the stagnant wall film or in both the film and bulk. During hypobaric storage the vapor mass fraction in an empty chamber is 0.46–0.56, it is even higher when plant matter is present, and since the flow is laminar and temperature difference between the commodity and wall is less than 2.5°C, mist formation might occur in the stagnant gas surface layers at the chamber roof or wall.

Any mist or fog formed during hypobaric storage would be in equilibrium with a supersaturated condition because the equilibrium vapor pressure over a convex mist or fog water droplet is higher than that over a plane surface of water:

$$e_r = e_\infty \exp[a/r] \tag{7.1}$$

where e_r (atm) is the saturation vapor pressure of a water droplet of radius r (μm), e_∞ is the saturation vapor pressure over a plane surface of water (atm), and $a \approx 3.3 \times 10^{-7}/T$ (meters). The fraction e_r/e_∞ increases with a decrease in droplet radius. In Table 7.1, $e_r/e_\infty =$ 1.0012 at a 10 μm radius corresponds to 0.12% supersaturation. This is the degree of supersaturation required for a 20-μm diameter (10 μm radius) wet fog[1] water droplet to be in equilibrium with the surrounding water vapor partial pressure. The droplet will evaporate if supersaturation is below the indicated value for the droplet radius and increase in size if supersaturation is higher than the indicated value. Both wet fog and 0.12% supersaturation would initially coexist in the LP storage chamber, and after this mixture reached a vacuum regulator operating at 25°C, flow to the vacuum pump would increase and stabilize when

Table 7.1 Fraction e_r/e_∞ for Different Sized Water Droplets at 0 C, Where r is the Radius of the Water Droplet (Tiedtke, 1987)

r (μm)	10^{-2}	10^{-1}	10	100
e_r/e_∞	1.128	1.012	1.0012	1.0001

In Public domain—Table 1.1 in Tiedtke (1987) entitled "Fraction e_r/e_∞ for different sizes of water droplets at 0° C."

[1]The international definition of fog is a visibility of less than 1 km (3300 ft); mist is a visibility of between 1 km (0.62 miles) and 2 km (1.2 miles); haze extends from 2 to 5 km (1.2–3.1 miles). Water droplets have a mean volume diameter of approximately 10-15 microns (μm) in dry fog; 20–30 μm in wet fog; and 60 μm in mist.

the storage chamber was no longer supersaturated and condensation ceased.

Water condensation has been problematic during hypobaric storage since the process was first described in 1966 (Burg and Burg). The box rows in Grumman/Dormavac hypobaric intermodal containers had to be stabilized with wood bracing because cardboard storage boxes were excessively weakened by exposure to or contact with condensed water. In laboratory systems, plant matter had to be placed above perforated desiccator plates or otherwise elevated to allow condensed water to drain and accumulate without flooding the tissue. The top boxes in a Vivafresh warehouse are covered with plywood panels, and in laboratory systems the commodity often has been wrapped with polyethylene film to prevent it from contacting condensed water. Surface water stimulates decay and may vacuum infiltrate into the commodity's intercellular system when the chamber is vented, injuring plant matter by clogging its intercellular air spaces. A vacuum regulator prevents these undesirable effects by preventing condensation.

Commodity weight loss during VacuFresh shipments can be decreased by substituting a vacuum regulator operating at $20-25°C$ in place of the vacuum breaker used in the original VacuFresh design (Section 5.3; Figure 5.1). The regulator causes the pumping speed to slow in response to the progressive decrease in metabolic heat production and transpiration that occurs during many weeks after harvest (Table 4.2). This prevents evaporative cooling from decreasing the commodities temperature and causing heat to radiate to the plant matter from the chamber wall. Eliminating access to environmental heat allows commodity weight loss to decline in proportion to the progressive decrease in metabolic heat production. A vacuum regulator also can maintain the relative humidity close to saturation in a Vivafresh hypobaric warehouse, preventing water from condensing and dripping onto boxes.

When a vacuum regulator operating at $20-25°C$ increases the rate at which supersaturated low-pressure air flows from the storage chamber to the vacuum pump, it reduces the available $[O_2]$ in the flowing air. The pressure must be controlled approximately 5 mmHg higher than the optimal value when a vacuum breaker is used to compensate for the decreased $[O_2]$ concentration (Burg, 2012).

CHAPTER 8

Low-Oxygen Injury

Exposing plant matter to hypoxia at atmospheric pressure induces the expression of genes, within 1 day, which increase the formation of alcohol dehydrogenase, pyruvate decarboxylase, and other enzymes and factors that promote fermentation. Excessively low $[O_2]$ causes toxic concentrations of ethanol and acetaldehyde to accumulate in plant matter (Fidler, 1951; Imahori et al., 2005; Table 8.2) and induce development of necrotic discolored tissues, off-odors, and other symptoms of low-$[O_2]$ injury, often within 1–4 days (Imahori et al., 2005; Ke et al., 1993, 1995). Hypoxia induces expression of the same genes during a hypobaric exposure, just as quickly (Paul et al., 2004), but normally this does not cause low-$[O_2]$ damage, and within 1 day after the plant matter is returned to atmospheric pressure it recovers its original genetic and physiological states, aerobic mode of metabolism, and fruits are able to ripen with normal flavor, aroma, and texture. If adequate incoming air changes are flowed at a hypobaric pressure to ensure the presence of at least 0.1% $[O_2]$ inside the storage chamber, toxic levels of ethanol and acetaldehyde do not accumulate and plant matter has a fresh odor when it is removed from LP. The optimal $[O_2]$ concentration for LP storage is 3- to 40-fold lower than the $[O_2]$-partial pressure causing low-$[O_2]$ injury at atmospheric pressure (Table 8.1). At 760 mmHg, 25°C, the respiratory inversion point occurs at 50–60% of the maximum respiration rate in peaches at 4% $[O_2]$, in apples at 4% $[O_2]$, in Japanese persimmon at 1% $[O_2]$, and in Satsuma mandurin at 1% $[O_2]$ (Kubo et al., 1996). At 23°C, the respiratory inversion point occurs at 2.5% $[O_2]$ in Bramley's Seedling apples (Figure 8.1, left). In LP, the inversion point in papaya at 10°C is 0.08% $[O_2]$ at 14 mmHg (Figure 8.1, right).

LP eliminates the $[O_2]$, $[CO_2]$, and $[C_2H_4]$ gradients that develop between the center and surface of a horticultural commodity's intercellular system in CA, MA, and NA at atmospheric pressure. These gradients arise due to gas production or consumption and the diffusive resistance of the commodity's peel and intercellular air spaces to O_2,

Hypobaric Storage in Food Industry. DOI: http://dx.doi.org/10.1016/B978-0-12-419962-0.00008-5

Table 8.1 The Highest Applied [O₂] Concentration Which Causes Low-[O₂] Injury to Plant Matter at Atmospheric Pressure, Compared to the Optimal % [O₂] for Hypobaric Storage

Commodity	CA	LP
	Injurious % [O₂]	Optimal % [O₂]
Green snap bean	<2	0.07
Asparagus	<5–10	0.30
Cauliflower	<2	0.13
Avocado	<2–3	0.10
Mushroom	<1	0.13
Papaya	<1–2	0.08
Mango	<3–5	0.10
Carambola	<5	0.25
Orange	<5	0.25

The percent [O₂] values indicated for LP are the lowest that have been tested and may not be the minimum which can be tolerated (Burg, 2004; Thompson, 2010).

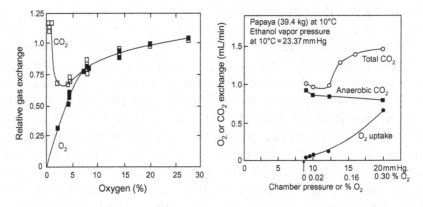

Figure 8.1 Relation between [O₂] concentration, O₂ uptake, and CO₂ production. Left: Influence of [O₂] concentration on O₂ uptake and CO₂ evolution by Bramley's Seedling apples at 23°C. The IP is 2.5% [O₂] (Street, 1963, originally published by Peragmon Press—now owned by Elsevier). STM permission for Elsevier's reuse of Figure 14.1. The influence of the percent oxygen upon the oxygen uptake and carbon dioxide evolution of Bramley's Seedling apples (Street, 1963). Right: O₂ uptake and CO₂ production by papaya stored in LP computed from the rate of entry of makeup air into the storage chamber and colorimetric gas detection tube measurements of [O₂] and [CO₂] present in atmospheric air and the vacuum pump exhaust. The IP is 0.08% [O₂] (Burg and Davenport, unpublished figure).

CO₂, and C₂H₄ transport through air-filled barriers (Burg, 1990, 2004; Hatton and Spalding, 1990; Leshuk and Saltveit, 1990; Dilley, 1978; Hatton et al., 1975; Lipton, 1975). LP enhances gaseous diffusion through air by as much as 100-fold (Figure 2.2), eliminating up to 99% of the surface to center [O₂] gradient that arises at atmospheric pressure. The respiratory inhibition caused by the ultra-low [O_2] present during LP storage reduces the gradient by an additional 10-fold (Burg and Kosson, 1983). Eliminating the [O_2] gradient causes the commodity's surface and center to be exposed to the same optimal low [O_2]-partial pressure for storage, whereas CA, MA, and NA are limited to higher applied [O_2] at the surface to avoid low-[O_2] damage in the commodity's center.

Because LP shifts the respiratory inversion point to a lower O_2-partial pressure, papayas do not experience low-[O_2] injury when they are stored at 15 mmHg, 10°C (0.16% [O_2]) for 3−4 weeks (Figure 8.1, right), and mangos are not damaged during 8 weeks at 13°C,15−20 mmHg (0.10−0.24% [O_2]) (Davenport et al., 2006). At atmospheric pressure and 13°C, mangos are injured by less than 3% [O_2]. At 10°C, less than 1.5−2% [O_2] eventually causes low-[O_2] injury to papayas (Thompson, 2010), and 0.2−0.4% [O_2] produces this disorder within 3 days (Yahia et al., 1992; Table 8.1). At 0°C and atmospheric pressure, chrysanthemum blooms are severely injured by 0.1% [O_2] within 1 day at 0.5°C (Ke and Kader, 1992), while at 10 mmHg (0.14% [O_2]) they are preserved without injury for 42 days (Burg, 2004). LP stores cold-intolerant avocado varieties at 13°C, 15−20 mmHg (0.10−0.24% [O_2]) for 49−60 days, without injury (Apelbaum et al., 1977c; Davenport and Burg, unpublished), but at atmospheric pressure they suffer low-[O_2] damage in 0.1−0.4% [O_2] within 1 day (Ke and Kader, 1992). Asparagus is injured by less than 5−10% [O_2] in CA, but has been stored in LP for 4−6 weeks without low-[O_2] injury in 0.42% [O_2], and likely tolerates 0.1% [O_2] (Dilley, 1977).

When a full load of 15°C carnation flowers was cooled in a Dormavac container, vacuum cooling began when the flower temperature approached 14°C and the pressure had decreased to 12 mmHg (Table 9.1, lower). At this "flash point," the partial pressure of water vapor equals the total chamber pressure, and the container's atmosphere is anaerobic. Because of the limited capacities of the container's refrigeration system and vacuum pump, air did not begin to enter until 5 h later when the flower temperature had decreased to 11.3°C (Table 9.1, upper) and the tank pressure reached the 10 mmHg set operational pressure. The carnations experienced low-[O_2] damage during this prolonged vacuum cooldown. Their leaves turned black several days later. This was the first time low-[O_2] damage was observed during an LP storage, and to avoid a recurrence Grumman specified that a commodity must be pre-cooled before it is loaded into a hypobaric system, and the pressure should be set at a value which guarantees the presence of at least 0.14% [O_2]. McKeown and Lougheed (1980), Salunkhe and Wu (1975), Tolle, (1972), Lougheed et al. (1977), and Kader (1975) speculated that any off-odors and off-flavors produced in LP might volatize and escape, and Wu et al. (1972) claimed that there is an inverse relationship between the hypobaric pressure and quantity of volatiles retained in tomato fruits stored at 12.8°C. Wu et al. attributed this result to a continuous evacuation of volatile organic compounds (VOCs) from the tomatoes, but their study was compromised by humidifying the incoming air at atmospheric pressure rather than a hypobaric pressure (Section 2.2) and supplying only 0.048 air changes per hour. This provided so little [O_2] that the tomatoes consumed most or all of the O_2 provided. Decreasing the pressure to 471 mmHg from the control ambient value of 646 mmHg in Logan, UT, where the experiment was performed, caused an average loss of 61.2% of tomato VOCs in 100 days, and at 102 mmHg the loss was 95.9%. A decrease in pressure to 471 mmHg hardly changes the diffusion rate of gases or vapors (Figure 2.2; Eq. (3.1)) and does not cause most biological vapors to boil at 12.8°C (Table 9.1), normally a modest pressure reduction has no effect on storage or ripening (Burg, 2004), and fruits stored at 10−20 mmHg ripen with normal flavor and aroma when they are removed from LP storage. These observations suggest that hypoxia rather than hypobaria caused the reduction in VOCs measured in Wu et al.'s tomato experiment. A study demonstrating that ultra-low [O_2] reduces the production of volatiles supports this explanation (Argenta et al., 2004).

Burg and Davenport studied the retention of fermentative VOCs and the development of low-[O_2] injury during LP and CA mango storage using methods described in Sections 5.5−5.7. Preclimacteric Keitt mangos were stored at 13°C, either at atmospheric pressure flowing one saturated incoming 0.145 ± 0.007% [O_2] air change per hour, measured by gas chromatography (GC), passing through an 11.1-L CA chamber containing 4.7 kg of fruit, or flowing one saturated 15 mmHg air change per hour, containing 0.105% [O_2], passing through a 170-L LP chamber containing 31.1 kg of fruit. On the third storage day, mangos in the CA chamber were emanating 21.7 ± 0.3 μmol/kg/h of ethanol and 1.3 ± 0.2 μmol/kg/h of acetaldehyde; in the LP chamber 57 ± 0.3 and 5.2 ± 0.2 μmol/kg/h of these VOCs, respectively. Similar rates of emanation were measured on the fourth day (Table 8.2, upper right), at which time air leaving the CA chamber contained $0.145 \pm 0.006\%$ [O_2] and $0.71 \pm 0.006\%$ [CO_2]. This indicated that none of the incoming [O_2] had been consumed and all CO_2 was produced by fermentation. Transpired commodity water caused the pressure regulator to increase the pumping speed and rate that air flowed from the LP chamber to the vacuum pump by 54% without changing the rate at which low-pressure 0.105% [O_2] entered the LP chamber (Section 5.3). This should have lowered the chamber [O_2] to 0.065%. A GC measurement on the fourth day indicated that $0.71 \pm 0.01\%$ [CO_2] was present in air entering the vacuum pump and $0.79 \pm 0.002\%$ [CO_2] in the vacuum pump's exhaust. The chamber [O_2] initially exceeded the concentration in the rarified air change due to the release of dissolved [O_2] from the mangos, and by the fourth day $0.078 \pm 0.002\%$ [O_2] was present in the air change. The LP fruit was not consuming O_2 and all CO_2 was produced by fermentation.

Metabolic CO_2 is produced in a gaseous state (Eqs. (4.14) and (4.15)) and does not require latent heat to escape from a horticultural commodity. Ethanol and acetaldehyde are produced as liquids

Table 8.2 Upper: Emanation Rate and Aqueous Concentrations of Ethanol, Acetaldehyde, and CO_2 in Keitt Mango Fruits After 4 Days' Storage at 13 C; Lower: The Apparent Membrane Permeability to Ethanol, Acetaldehyde, and CO_2 Calculated as mmol/kg/h of Gas or Vapor Emitted from the Fruit Divided by the Gas or Vapor's mmol/kg Concentration in the Tissue

	Concentration in Fruit's Aqueous Phase (mmol/kg)		Emanation Rate (μmol/kg/h)	
	CA	LP	CA	LP
C_2H_5OH	1.89 ± 0.14	2.09 ± 0.14	24.9 ± 1.4	63.0 ± 6.0
CH_3CHO	0.22 ± 0.015	0.32 ± 0.047	1.5 ± 0.2	4.6 ± 0.7
CO_2	11.7 ± 2.1	80 ± 2.2	799 ± 73	1210 ± 85

	(mmol/kg/h)/(mmol/kg)		
	CA	LP	CA/LP
C_2H_5OH	75.9 ± 5.5	33.2 ± 6.5	2.3
CH_3CHO	146.6 ± 9.9	69.5 ± 10.2	2.1
CO_2	1.46 ± 0.26	0.66 ± 0.18	2.2

Upper: The aqueous concentrations of these substances in each fruit were determined by GC, and results from four fruits are averaged. Measurements of dissolved tissue [CO_2] did not include CO_2 released from bicarbonate and carbonate since the mango pH was less than 5.0 (Table 5.2). Six untreated mangos initially contained 23 ± 4 μM ethanol, 75 ± 9 μM acetaldehyde, and 12.4 ± 1.0 mM CO_2 per kg of tissue. Lower: Both in CA and LP, the mmol/kg/h emanated per mmol/kg in the tissue was in the ratio 100:52:1 for acetaldehyde, ethanol, and CO_2, respectively.

(Eq. (4.16)) and must acquire sufficient latent energy to change state from liquid to vapor before they can evaporate or boil from a horticultural commodity.

Thermistor temperature measurements indicated that the mangos were warmer than the chamber wall and not receiving environmental heat, so the only source of latent energy was fermentative heat. The CO_2 emanation rates compiled in Table 8.2 (upper right) indicate that fermentative heat was being produced at a rate of 15.6 cal/kg/h in CA and 24.2 cal/kg/h in LP (Eq. (4.17)). To vaporize 1 g of liquid water at 13°C requires 588.6 cal of latent heat. Therefore, the fermentative heat generated in the LP chamber was capable of vaporizing 0.041 g of transpired water per hour (Eq. (4.17)). If the ethanol and acetaldehyde dissolved in 0.041 g of the fruit's water was released each hour by transpiration (Table 8.2, upper left), this would only account for 0.14% of the total ethanol and acetaldehyde emanating from the mangos. These VOCs were not escaping in transpired water.

Fermentation produces 1 mol of CO_2 accompanying each combined mole of ethanol and acetaldehyde, but Table 8.2 indicates that the ethanol + acetaldehyde present in the mangos and emanating from them accounts for less than 5% of their production. Likewise, under anaerobic conditions, pepper fruits produce 5 μL/g/h of CO_2 while emanating only 2.5 nL/g/h of acetaldehyde and 9.4 nL/g/h of ethanol (Zuckerman et al., 1997). In both instances, the ethanol and acetaldehyde emanation rates are vastly lower than their production rates (Eq. (4.17)), indicating that most of the pyruvate produced by glycolysis has been reduced to lactate, and/or acetaldehyde and ethanol are being consumed or metabolically altered. Apples rapidly metabolize ethanol when O_2 is scarce (Knee and Hatfield, 1981), and under anaerobic conditions use ethanol to form large amounts of ethyl acetate, and lesser quantities of ethyl propionate, ethyl butyrate, ethyl-2-methyl butyrate, ethyl hexanoate, ethyl heptanoate, and ethyl octonoate (Mattheis et al., 1991). In 0.25% [O_2], avocados reduce pyruvate to lactate and ferment pyruvate to acetaldehyde and ethanol, giving rise to 5 mM lactate, 0.35 mM acetaldehyde, and 15 mM ethanol in the tissue (Ke et al., 1995). In the absence of O_2, potato tubers produce lactic acid and very little ethanol (Barker and el-Saifi, 1953).

Since acetaldehyde and ethanol do not escape from mangos dissolved in transpired water, they must pass through the plasmalemma and cell wall, vaporize into and diffuse through the intercellular system, and exit through mango lenticles into the air change. The gradient that causes a small lipid soluble nonpolar molecule to diffuse through the lipid portion of the plasma membrane is the difference between the molecule's aqueous concentration at the inner and outer surface of the membrane. Because lipid membranes and aqueous solutions have different solvent properties, a partition coefficient (K) is applied to estimate the concentration gradient moving solute molecules through the membrane lipid (Section 5.1). The dimensionless partition coefficient is defined as the ratio between a dissolved solutes concentration in the membrane's lipid material and that in an adjacent aqueous phase. It is estimated for each solute by measuring the ratio between the solute's concentration in a lipid material such as olive oil or octanol, which mimics membrane lipid, and the solute's equilibrium concentration in water. This convention is based on the plasma membrane's high lipid content and experiments showing that the relative ease with which small nonpolar molecules passively penetrate through the plasmalemma often is correlated with their lipid solubility and molecular size (Collander, 1937). The ethanol concentration at the outer surface of the plasma membrane, estimated by multiplying the ethanol concentration in the tissue (Table 8.2) by ethanol's 0.03 olive oil/water (Collander, 1937) or octanol/water (Sangster, 1997) partition coefficient, predicts that if the amount of liquid ethanol that theoretically should diffuse to the outer surface of the plasmalemma continuously vaporized from it, a concentration of 198.4 ppm of ethanol vapor would be maintained in the CA air change and 171 ppm in the LP air change. These values are closely similar to the 236.7 and 171 ppm ethanol concentrations measured in air flowing through the CA and LP chambers, respectively.

Acetaldehyde's high polarity (2.7 debye) elevates its boiling point (Table 9.1) and alters its membrane permeability by increasing its water solubility and ability to H-bond water. CO_2 is able to pass through the same aquapores in the plasma membrane that transfer liquid water (Maurel et al., 2009). These processes alter the rate at which CO_2 and acetaldehyde permeate the plasma membrane.

Both in LP and CA, low-[O_2] damage occurs when the [O_2] concentration is less than that present at the RQ inversion point (Figure 8.1), but in LP the RQ inversion point arises at a much lower [O_2]

concentration. At atmospheric pressure, low-[O_2] injury develops when mangos are exposed to less than 3% [O_2], but not in LP when only 0.1% [O_2] is available (Table 8.1). In the mango test described in Table 8.1, the dense commodity load, a low incoming air-change rate, and accelerated flow of low-pressure air from the storage chamber to a vacuum regulator operating at 25°C caused low-[O_2] injury by reducing the steady-state [O_2] concentration in the hypobaric chamber to 0.065%. In LP, low-[O_2] damage develops in papayas at 0.02 but not 0.08% [O_2] (Figure 8.1, right).

CHAPTER 9

Pervaporation

9.1 COMMERCIAL PERVAPORATION

Pervaporation is a method used in commercial industry, recently under intensive development to remove water from ethanol biofuels. It is unique among membrane separation processes in including a phase change from liquid to vapor. A synthetic, very strong membrane separates a solution containing one or more volatile liquids, usually at atmospheric pressure, on one side, from a vacuum condition, on the other side of the membrane. Liquid volatiles pass through the membrane at rates determined by their specific nature and the characteristics of the membrane. Heat is supplied to vaporize the permeating volatile liquid(s) at the evacuated side of the membrane, and to maximize mass transport across the membrane the vapor pressure of a component on the permeate side is kept low by evacuating (*vacuum pervaporation*) or purging the permeate (*sweep gas pervaporation*).

Pervaporation has three steps: the sorption of permeate at the solution feed/membrane interface, diffusion across the membrane due to a chemical potential gradient (rate-determining step), and desorption into a vapor phase at the permeate side of the membrane. The first two steps are primarily responsible for the perm selectivity, and because phase change occurs during the third step, the membrane temperature and pressure gradient have significant effects on the separation performance. To ensure that the permeate boils at the vacuum side of the membrane, an amount of energy must be supplied that is at least as great as the permeate's heat of vaporization, and the permeate's pressure must be kept lower than its saturation vapor pressure at the process temperature. The separation is based on physical–chemical interactions between the membrane material and the permeating molecules, independent of vapor/liquid equilibrium. The driving force for the mass transfer of permeants from the membrane's feed to the permeate side is the volatile organic compound (VOC)'s chemical potential or partial pressure gradient across the membrane, not the

Hypobaric Storage in Food Industry. DOI: http://dx.doi.org/10.1016/B978-0-12-419962-0.00009-7

VOC's volatility. Raising the feed pressure increases the chemical potential gradient and flux through a pervaporation membrane, 10-fold for a feed pressure elevation from 1 to 10 atm. Decreasing the pressure on the "permeate" side of the membrane also increases the permeation rate of a feed component.

9.2 PERVAPORATION DURING HYPOBARIC STORAGE

At atmospheric pressure, the boiling points of pure ethanol, acetaldehyde, and water are 78.4°C, 20.2°C, and 100°C, respectively (Table 9.1), but water and a dilute aqueous ethanol solution boil at the same temperature regardless of the pressure, and a turgid plant cell's 4−20 atm hydrostatic pressure prevents cytoplasmic water and dissolved ethanol or acetaldehyde from boiling at a physiological temperature. Palta and Stadelmann (1980) found that C^{14}-ethanol and H^3-water cross the plasma membrane independent of each other and do not interact in the presence or absence of a simultaneous influx of water as they passively efflux through epidermal cells in *Allium cepa* bulb scales. This suggests that when a plant cell's high hydrostatic pressure accelerates the diffusive efflux of a small fat-soluble molecule such as ethanol through the hydrophobic lipid portion of the plasma membrane (Collander, 1937), the ethanol moves independent of water, and upon reaching the membrane's outer permeate surface it can acquire metabolic or sensible heat and evaporate separated from water. During LP storage, a horticultural commodity's cell walls and plasma membranes withstand the differential force exerted by a vacuum, on one side, and the cell's hydrostatic pressure on the other side. When ethanol or acetaldehyde efflux to the outer surface of the plasma membrane independent of water, instead of evaporating they acquire sensible or metabolic heat and boil at the temperature and pressure combination indicated in Table 9.1.

Lowering the pressure to 15 mmHg at 13°C caused a 2.1- to 2.3-fold acceleration in the rate at which acetaldehyde, CO_2, and ethanol diffused through the plasma membrane of mangos into their intercellular system (Table 8.2, *lower*; CA/LP). A preclimacteric mango cell's turgor pressure is at least 4 atm, and since hypobaric storage only adds ≤0.994 atm to the pressure gradient across the plasma membrane, it would only increase the efflux of these molecules by at most 20%. Therefore, the 2.1- to 2.3-fold greater loss of ethanol,

Table 9.1 (*Upper*) Boiling Point and (*Lower*) Saturated Vapor Pressure of Water, Ethanol, Acetaldehyde, and Ethyl Acetate at Various Storage Temperatures and Pressures

Storage Pressure (mmHg)	Boiling Point (°C)			
	Water	Ethanol	Acetaldehyde	Ethyl Acetate
760	100	78.4	20.2	77.0
80	47.1	27.1	− 26	49.1
70	44.5	31	− 29	48.7
60	41.6	26	− 31.4	48.3
50	38.2	23	− 34	47.9
40	34.5	19	− 37.8	47.5
30	29.0	14	− 42	47.1
20	22.2	8.0	− 47.8	46.7
15	17.6	4.0	− 53	46.5
10	11.3	− 2.3	− 56.8	46.3
Temperature (°C)	Saturated Vapor Pressure (mmHg)			
0	4.6	11.8	330	24.6
1	4.9	12.6	348	26.1
2	5.3	13.6	373	27.7
3	5.6	14.6	384	29.3
4	6.1	15.6	400	31.1
5	6.5	16.7	415	33.0
6	7.0	17.9	431	34.9
7	7.5	19.1	451	37.0
8	8.0	20.5	469	39.2
9	8.6	21.9	488	41.4
10	9.2	23.4	506	43.8
11	9.8	25.0	523	46.3
12	10.5	26.6	538	48.9
13	11.2	28.4	570	51.5
14	12.0	30.3	592	54.5
15	12.8	32.2	616	57.4
16	13.6	34.3	640	60.5
17	14.5	36.4		63.4
18	15.4	38.7		66.7
19	16.4	41.1		70.2
20	17.5			
21	18.6			
30	31.8			

acetaldehyde, and CO_2 from mangos during LP storage must be mainly due to their increased rate of diffusion through the intercellular system (Figure 2.2; Eq. (3.1)), and to the establishment of a condition favoring boiling in lieu of evaporation at the outer surface of the plasma membrane (Table 9.1). An increase in diffusivity would affect all gases and vapors to the same extent.

CHAPTER 10

Capillary Condensation in Non-Waxed Cardboard Boxes

Throughout 3 days, the humidity could not be elevated above 92% in a laboratory hypobaric storage chamber containing an empty nonwaxed cardboard box when one 13°C, 20 mmHg, saturated air change per hour was flowed and the pressure was controlled by a vacuum breaker. After the chamber was vented and the box removed, the humidity rapidly reached saturation when the chamber was reevacuated and returned to the same operating condition. The cardboard was lowering the humidity. This same behavior occurred in a Vivafresh hypobaric warehouse filled with 7207 kg of roses stored in nonwaxed cardboard boxes. The pressure was controlled at 11.1 mmHg by a vacuum breaker, and the warehouse was mechanically humidified by two air changes per hour of low-pressure air saturated at the storage temperature. Ten air changes should have brought the humidity in an empty warehouse to 99.9% in 5 h. Instead, filled with roses packed in nonwaxed cardboard boxes, it took 12.5 days for the chamber humidity to reach that level even though mechanical humidification with sufficient moisture to saturate the air was supplemented with transpired moisture from the roses (Figure 10.1).

Cardboard is a capillary-porous colloidal cellulosic material containing randomly distributed micropores, macropores, and amorphous regions of diverse shapes and sizes. It is highly hygroscopic, absorbing large amounts of adsorption-bonded moisture because its polymeric molecular chains contain hydroxyl groups. Kuts et al. (1975) studied the influence of a vacuum and atmospheric pressure on the internal mass transfer of water vapor in electrical-grade cellulosics, including cardboards used for high-voltage insulation. Moisture diffused through cardboard micro- and macropores 20- to 50-fold faster in a vacuum than at atmospheric pressure, and pure water vapor was absorbed 50- to 150-fold more rapidly. After an initial steep rise, the rate of moisture absorption under vacuum rapidly decreased and approached an end point equilibrium.

Hypobaric Storage in Food Industry. DOI: http://dx.doi.org/10.1016/B978-0-12-419962-0.00010-3

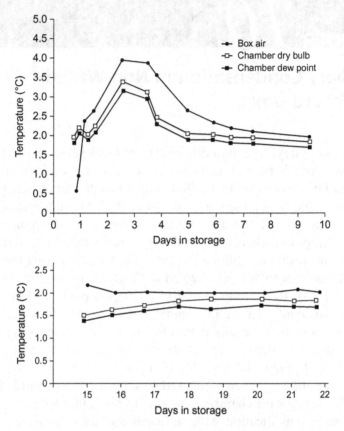

Figure 10.1 Temperature changes in a hypobaric warehouse measured with ± 0.05° C thermistor probes inserted into the chamber air and into 220 g cardboard boxes (16 × 25 × 55 cm³) containing 8 bunches (6320 g) of roses. Two air changes per hour of 11.1 mmHg saturated air was flowed through the chamber each hour. Upper: after initial pump-down. Lower: after second pump-down at 14 days.

The Kelvin equation predicts that surface tension (σ) causes the saturation vapor pressure (e) over the concave liquid surfaces in the interstices between a cardboard box's hydrophilic cellulose microfibrils to be higher than the saturation vapor pressure ($e°$) over a large plane surface:

$$\ln e/e° = -2\sigma V/rRT \qquad (10.1)$$

where V is the molar volume, r is the radius of curvature of the interstices, T is the temperature, and R is the universal gas constant. In cardboard, r varies between 25 and 50 μm in macrocapillaries and between 1.25 and 2.5 μm in microcapillaries (Schneider et al., 1986). The increased surface tension causes water vapor to condense in these interstices at a higher temperature than is needed to change water's physical state above a plane surface. A low-pressure also increases the convective coefficient for film condensation in the microfibrilar spaces (Section 4.1; Özisik, 1985) and hastens the transport of water vapor from the flowers to the cardboard by accelerating the diffusion of moisture through the air/water mixture contained in the box (Figure 2.2; Eq. (3.1)).

Dry wooden field crates increase in weight by 6.7—8.0% at atmospheric pressure during an apple storage season, and corrugated fiberboard containers also take up moisture (Hardenburg et al., 1986). This same behavior occurs during hypobaric storage in nonwaxed cardboard boxes, but the increase in box weight is larger and occurs much more rapidly in a vacuum. Water condensed in the cardboard and increased its weight by 20.1% in 6 weeks when 120 nonwaxed cardboard boxes, each containing 6320 g (8 bunches) of Freedom or Charlotte roses, were stored in a Vivafresh hypobaric warehouse. Regardless of whether the boxes were empty or filled with roses, in 4 weeks a saturated storage humidity caused an 18.4—18.8% weight increase in the cardboard. By the fourteenth day, each cardboard box (specific heat = 0.44 cal/g °C) had condensed 195.2 g of water and released 594 cal of latent heat per gram of condensed water, a total of 115.2 kcal. This amount of heat was capable of raising the cardboard's temperature by 213°C if it was not removed. Most of the heat was transferred to the flowers since the entire box surface area moving heat inward by radiation and convection was 22.8-fold larger than the area radiating heat from one end of the box to the warehouse wall and 11.4-fold larger than the area transferring heat by outward convection from both ends of the box into the warehouse air. Heat transferred to the roses warmed them from 0.6°C to 3.9°C within a few days after the warehouse was initially evacuated (Figure 10.1, upper), but after the warehouse was vented and again pumped down on the fourteenth day, no significant temperature rise occurred in the flowers because water condensation in the cardboard had nearly ceased (Figure 10.1, lower). Measurements of flower weight loss and cardboard weight gain, and visual examinations made when the chamber was vented and opened, indicated that during the initial 17 days water did not condense in the warehouse. After 17 days, the humidity became supersaturated, and due to transpired water vapor's high buoyancy (Eqs. (4.17) and (5.1)), it rose to the roof, where it condensed, and then dripped back onto the boxes.

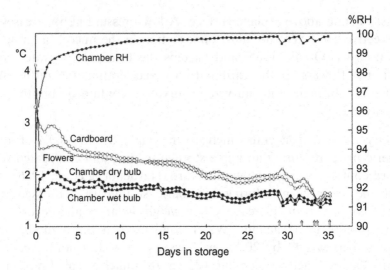

Figure 10.2 Cardboard box and flower temperature, warehouse air temperature, and dew point and RH measurements during hypobaric storage of a full load of nonwaxed cardboard boxes containing a total of 7207 kg of Freedom red roses. Two saturated 11.1 mmHg, 2°C air changes were flowed through the warehouse each hour. Arrows indicate times when the chamber was vented to atmospheric pressure for 2–4 h in order to remove partial loads.

In a subsequent storage, a ± 0.05°C thermistor probe was imbedded in the cardboard of a $25 \times 16 \times 55\,\mathrm{cm}^3$ box, a second probe was inserted among the 12 flower bunches contained in the box, and the warehouse was fully loaded with nonwaxed cardboard boxes containing Freedom roses (Figure 10.2). During pump-down, the chamber relative humidity (RH) decreased to 93% and the temperature of the flowers (specific heat = 0.87 kcal/kg °C) declined to 2.5°C from an initial value of 4.2°C. By the second day, heat generated by water condensation in the cardboard had warmed the cardboard to 4°C and the roses to 2.9°C, while the chamber RH rose to 98.5%. As the flowers were colder than the cardboard during the initial 12 days (Figure 10.2), they received heat from the cardboard by radiation, convection, and conduction during that time, and emitted this heat as well as metabolic heat by evaporative cooling, the only heat transfer mode available to the flowers in the absence of a positive outward temperature gradient. This behavior continued for the initial 12.5 days, until the rate of heat production due to water condensation in the cardboard had slowed to such an extent that metabolic heat caused the flower temperature to equal the cardboard temperature (Figure 10.2, 12.5 days). The chamber RH then reached supersaturation, and transpired commodity water began condensing on the chamber roof. Subsequently, a positive outward temperature gradient developed and heat began to be transferred by radiation and convection from the flowers to the cardboard, rather than in the reverse direction, while the small amount of heat being generated by water condensing in the cardboard was transferred by radiation and convection from the cardboard to the chamber wall and air. Because the flowers now were the warmest objects in the storage chamber, they could no longer acquire heat from the cardboard or any other environmental source, and only their metabolic heat needed to be removed by evaporative cooling.

Condensed water increased the weight of each $16 \times 25 \times 55$ cm^3 nonwaxed cardboard box by 18% in 5 weeks, generating 138.9 kcal of latent heat. This heat was transferred to the roses by radiation (Eq. (4.9)) and convection (Eqs. (4.1) and (4.2)) from the entire box surface area.[1] Radiation also transferred heat from one end of each box to the adjacent wall of the hypobaric warehouse, and convection from two ends of the boxes into the chamber air. In 35 days, 156.7 kcal of heat generated by water condensation in the cardboard was transferred to the chamber wall, and 123.3 kcal to the flowers. This caused a 207.6 kg weight loss from the flowers when they dispelled the acquired heat by evaporative cooling. The measured flower weight loss was 380.1 kg in 35 days. Heat generated by water condensing in the cardboard caused 55% of the total weight loss, metabolic heat caused the remainder of the weight loss. The heat transfer calculation ignores conduction from the cardboard to the outermost flowers because that cannot be accurately computed, and likely is insignificant since it is limited by the small area of random contact between the surfaces and the high thermal contact resistance caused by a vacuum (Section 4.4).

During the storage described in Figure 10.2, flower bunches located adjacent to the upper and lower box surfaces received direct radiation from 4100 cm^2 of cardboard surface, those in the center of the box from 675 cm^2, so each bunch located at the upper and lower surfaces received three times as much radiant heat as a bunch located in the center of the box. This caused the exterior bunches to lose the most

---

[1]Assumptions: cardboard emissivity = 0.81; rose leaf emissivity = 0.96; chamber wall emissivity = 0.55; convective heat transfer coefficient at 11.1 mmHg = 0.196 kcal/h m^2 °C; warehouse wall temperature = warehouse air dry bulb temperature (Burg and Kosson, 1982, 1983; Table 4.1).

water (Table 10.1). The data indicate that both with and without Mylar® the weight loss per bunch is not directly proportional to the amount of radiant and convective heat transferred to the flowers. Instead, before evaporative cooling removes the acquired heat, a portion is redistributed by convection and secondary radiation from upper and lower flower bunches to central bunches.

A Mylar® liner inserted in a nonwaxed cardboard box (Table 10.1, lower) reduced flower weight loss from 384.95 (6.78%) to 252.15 g (4.51%) during 35 days, The emissivities of plant matter, cardboard, and the shiny side of Mylar® are 0.96, 0.81, and 0.2, respectively (Table 4.1). These values inserted in Eq. (4.10) indicate that Mylar® should be approximately 88% effective in blocking radiation from cardboard to the plant matter. The actual efficiency blocking weight loss was less than 88% because the emissivity of the nonshiny backside of the Mylar® is less than 0.2, and approximately 5.8% of the heat transfer from cardboard to the plant matter was due to convection (Burg and Kosson, 1982, 1983). The Mylar® was at least 69.6% effective blocking weight loss. The evaporative weight loss caused by metabolic heat was less than 2.83% during 35 days (Table 10.1, lower).

The temperature difference between roses and the air dry bulb (Figure 10.2) decreased from 0.5°C to 0.25°C between the 12.5th and 28th day of storage. This indicates that metabolic heat production slowed by 50% during that interval, in agreement with data in Table 4.2 showing that heat production by flowers decreases by an average of 54.2% during the first 20−24 days of storage. Between the fourth and sixth week (Figure 10.2), when the flowers no longer were receiving heat from the cardboard, their rate of weight loss was 0.05% per day. Therefore, initially they were losing slightly more than 0.1% of their weight per day due to metabolic heat. In laboratory studies with "naked" Madam Delbar rose bunches at 10 mmHg respiratory heat accounted for 94.3% of their weight loss, with CO_2 emanation responsible for the remaining 5.7%. During the 35-day hypobaric storage illustrated in Figure 10.2, the roses should have only lost 2.5% of their weight due to metabolic heat. The 2.83% average weight loss in 35 days from central bunches shielded from radiation by Mylar® (Table 10.1, lower) was only slightly higher than this value.

To verify that heat generated by condensation in nonwaxed cardboard boxes caused a substantial part of the flower weight loss

Table 10.1 Weight Loss (g) During the 35 Days of Storage Illustrated in Figure 12.5, from Rose Bunches Situated Sequentially Between the Top and Bottom Surfaces of $16 \times 25 \times 55$ cm^3 Nonwaxed Cardboard Boxes

No Mylar	Initial	35 Days	Change	% Weight Change
Cardboard	1298.30	1532.10	233.80	**+18.00**
Bunch #1	460.23	425.07	35.16	−7.64
Bunch #2	466.97	435.07	31.90	−6.83
Bunch #3	471.77	439.67	32.10	−6.80
Bunch #4	473.86	442.27	31.59	−6.67
Bunch #5	504.97	480.07	24.90	−4.93
Bunch #6	476.53	459.00	17.53	−3.68
Bunch #7	477.87	450.46	27.41	−5.74
Bunch #8	483.87	452.97	30.90	−6.39
Bunch #9	484.07	444.73	39.31	−8.12
Bunch #10	448.20	415.60	32.60	−7.27
Bunch #11	459.07	422.37	36.70	−7.99
Bunch #12	479.65	434.80	44.85	−9.35
Bunches total	5227.99	4879.71	384.95	**−6.78 ave.**
Mylar®	**Initial**	**35 Days**	**Change**	**% Weight Change**
Cardboard	1310.55	1531.60	221.05	**+16.96**
Bunch #1	442.55	417.85	24.70	−5.58
Bunch #2	507.95	484.40	23.55	−4.64
Bunch #3	449.25	423.05	26.20	−5.83
Bunch #4	443.50	423.45	16.80	−3.79
Bunch #5	454.95	439.80	15.15	−3.33
Bunch #6	450.55	440.80	9.75	−2.16
Bunch #7	496.60	480.65	15.95	−3.21
Bunch #8	432.35	421.10	11.25	−2.60
Bunch #9	467.35	437.10	30.25	−6.47
Bunch #10	494.35	467.55	26.80	−5.42
Bunch #11	501.35	472.75	28.60	−5.70
Bunch #12	431.70	408.55	23.15	−5.36
Bunches total	5572.45	5317.05	252.15	**−4.51 ave.**

The flowers were packed as four bunches of upper layer (#1 to #4), middle layer (#5 to #8), and lower layer (#9 to #12) in the box. Upper: average from three boxes. Lower: average from two boxes with Mylar® slip sheets inserted between the cardboard and flowers to reflect radiation.[2]

[2]The cost of a 1 mil Mylar® slip sheet inserted in a rose box is less than $1.75 in a 50-cm wide box and less than $0.90 in a 25-cm wide box.

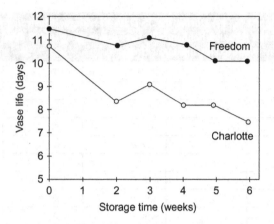

Figure 10.3 Vase life of Freedom and Charlotte roses, packed in nonwaxed cardboard boxes, after various storage times in a Vivafresh hypobaric warehouse. A result similar to that for Freedom roses was obtained with Alstroemeria *flowers during a 7-week storage.*

during the hypobaric storage depicted in Figure 10.3, a full-load storage was performed in which 12 bunches of Freedom roses were included, packed as upper, middle, and lower tiers in plastic boxes having the same dimensions as the nonwaxed cardboard boxes used to store the rest of the flowers. The average weight loss in nonwaxed cardboard boxes during 19 days was 5%, and in plastic boxes only 1.57%, including the 0.24% weight loss caused by vacuum cooling during the initial pump-down. The entire weight loss caused by water condensation in cardboard was eliminated in plastic boxes, and the flowers did not experience an increase in temperature after the initial pump-down was completed. All previous horticultural commodity shipments in hypobaric intermodal containers, and storages in hypobaric warehouses, were carried out with commodity packed in nonwaxed cardboard boxes. Cardboard cartons were too large to be used in laboratory chambers, and this caused results and weight loss comparisons obtained in laboratory studies, hypobaric warehouses, and intermodal containers to differ significantly.

When roses in plastic boxes were kept for 40 days in leak-free 170-L aluminum laboratory vacuum chambers at 10 mmHg, 0°C, flowing two saturated low-pressure air changes per hour, flowers wrapped with nonwaxed cardboard sleeves lost 15% of their water and the cardboard's weight increased by 18%, while flowers wrapped with a thin paper/plastic sleeve only lost 5% of their weight.

Senescence-associated genes (SAGs) induced by water stress promote senescence and petal wilting in all flower types. This shortens vase life, even though the flower is able to recover full turgor when it is refreshed in water (Thomas, 1980; Wagstaff et al., 2010). Freedom and Charlotte roses lost 1.4 and 2.7 days of vase life, respectively, during a 6-week Vivafresh storage in nonwaxed cardboard boxes (Figure 10.3). Half of the loss in Freedom roses, and 3/4 of the loss in Charlotte roses, occurred during the first 2 weeks while the flowers were water-stressed and transpiring at an excessive rate in response to heat generated by water condensation in the nonwaxed cardboard storage boxes. After 2 weeks, when heat no longer was being transferred from cardboard to the roses, they experienced a partial recovery of vase life.

Paul et al. (2004) studied gene changes in *Arabidopsis* seedlings grown on the vertical surface of nutrient agar plates incubated inside a light growth chamber operated at 85% RH, 22°C, either at 75 mmHg or with 2% or 21% [O_2] present at atmospheric pressure. Transpiration raised the RH to 95% or higher inside petri dishes in the 75 mmHg chamber, in which the [O_2] concentration was 1.5% due to the presence of approximately 18.8 mmHg of water vapor and 57 mmHg of 22°C air. During 24 h, more than 200 genes were differentially expressed in seedlings exposed to 75 mmHg, and this response was markedly different from that to 2.0% [O_2] at atmospheric pressure. Abscisic acid related drought-induced pathways were engaged at 75 mmHg, these same genes were repressed or unaffected by 2% [O_2] at atmospheric pressure, and genes for fermentative pathways were induced by both 1.5% [O_2] at 75 mmHg and 2% [O_2] at atmospheric pressure. Paul et al. concluded that the response of *Arabidopsis* to a 75 mmHg pressure is "a specific adaptation to perceived desiccation that is overlaid upon an adaptation to low [O_2]."

Photon absorption in a light growth chamber is linearly related to the light intensity. Photon utilization for CO_2 fixation during photosynthesis exhibits saturation kinetics, and photon absorption in excess of the capacity of photosynthesis is released as heat (Bowyer and Leegood, 1997). In Paul et al.'s study, the light growth chamber's 400-W metal halide lamp provided 400−700 nm radiation, of which 30% reaching a petri dish would be incompletely absorbed, 24% would be lost degrading short wavelength photons to 700 nm energy, and

68% of the energy absorbed by the *Arabidopsis* would not be incorpo-
rated in glucose (Hall and Rao, 1999). As most of the incident light
energy is converted to heat, the *Arabidopsis* seedlings transpired excessive water at 75 mmHg for the same reason that roses lose extra water
responding to heat received when water vapor condenses in nonwaxed
cardboard boxes.

Floral weight loss must be kept below 10% to avoid senescence and
a large decrease in vase life (Agrotechnology and Food Sciences
Group—Wageningen). Freedom roses stored in nonwaxed cardboard
boxes in a Vivafresh hypobaric warehouse had an average weight loss
of 6.78% in 35 days and still retained excellent vase life (Figure 10.3).

Heat is generated by capillary condensation in nonwaxed cardboard
boxes mainly during the first few weeks of exposure to a saturated
humidity. This is a major portion of the time required for a typical sea
shipment. In a VacuFresh hypobaric intermodal sea container, all of
the chamber moisture originates from the plant matter's transpiration.
Therefore, water condensation in nonwaxed cardboard boxes lowers
the chamber RH to a greater extent in VacuFresh than in Dormavac
containers or Vivafresh warehouses where transpired water is supplemented with mechanical humidification. Nonwaxed cardboard boxes
should not be used in VacuFresh containers.

CHAPTER 11

Insect Quarantine

The United States and other countries have regulations preventing importation of horticultural commodities that have not been treated by an approved method ensuring the mortality of 99.997% of specified insect pests. Fumigation with ethylene dibromide, carbon disulfide, hydrogen cyanide, and methyl bromide previously satisfied this requirement, but ethylene dibromide was banned in the United States and other countries due to suspected carcinogenicity (Wills et al., 1989) and the use of carbon disulfide and hydrogen cyanide has been discontinued because high toxicity and flammability make these substances hazardous to workers and potentially dangerous to the environment. The quarantine problem became acute when methyl bromide, the fumigant of choice, was scheduled to be discontinued. Public concern about chemical residues, and the damaging effects that insecticides pose to many fruits and vegetables had already prompted a shift to high- or low-temperature exposures as alternative procedures for killing insects; however, temperature extremes are injurious to many horticultural commodities.

CA and LP are under investigation as possible alternative chemical-free quarantine treatments. At an elevated temperature, CA kills many types of insects within a few days, but simultaneously damages most horticultural commodities (Mitcham et al., 2006). If insects are reliably killed by the low [O_2] concentration present at the optimal pressure for low-temperature hypobaric storage, then LP could satisfy the quarantine requirement at no additional expense during a prolonged in-transit exposure.

The cause of insect mortality at a low pressure has been debated for more than 50 years. Experiments with *Aedes aegypti* adults and housefly pupae suggested that "mortality of mosquitoes at very low pressures might stem from at least three factors: dehydration, a lack of O_2, and a low pressure *per se*" (Galun and Fraenkel, 1961, referred to in Navarro, 1978).

Hypobaric Storage in Food Industry. DOI: http://dx.doi.org/10.1016/B978-0-12-419962-0.00011-5

11.1 LETHAL EFFECT OF A LOW HUMIDITY

Decreasing the relative humidity (RH) in a chamber kills insects (Narayanan and Bhambhani, 1956; Munro, 1959; Sharplin and Bhambhani, 1963; Thornton and Sullivan, 1964; El Nahal, 1953; Chen et al., 2005b, 2006, 2008; Jay et al., 1971; Jay and Cuff, 1981). During exposure to both low [O_2] or high [CO_2] in controlled atmosphere storages (Pearman and Jay, 1970; Jay et al., 1971) and in LP (Zhang et al., 2005a; Calderon and Navarro, 1968), weight loss and mortality increase in parallel when the RH is decreased. Dehydration, with weight losses varying from 25% to 60%, killed larvae of *Hylotrupes bajulus* (old house borer) after they were inserted into wood of various moisture contents and kept at 20 mmHg, 20°C (Chen et al., 2005b, 2006). Adults, larvae, and eggs of *Anoplphora glabripennis* (Asian long-horned beetle) and *Agrilus planipennis* (emerald ash borer) were killed at weight losses of 26−40% when they were exposed to vacuum directly or inserted into wood at various moisture levels. Lethal vacuum time was directly related to larval weight loss (Chen et al., 2008). When insects oviposit their eggs *inside* stored plant matter, the life stages which develop are protected from desiccation by the saturated atmosphere present inside the host's intercellular system and are not reliably killed by a low pressure even though the stored plant matter's respiration causes the [O_2] to be lower inside the commodity than it is in the storage atmosphere.

Researchers have grossly overestimated the vapor pressure gradient required to rapidly desiccate an insect life stage during LP storage, especially when the insect is present on a horticultural commodity's surface receiving respiratory heat from the host (Calderon et al., 1966; Calderon and Navarro, 1968; Navarro and Calderon, 1978, 1979; Navarro, 1974, 1975; Navarro and Donahaye, 1972). The surface to volume ratio of a 10-cm diameter spherical fruit is 100-fold smaller than that of a 1-mm diameter insect life stage. Therefore, an insect of that size will tend to evaporate half its water 100-fold faster than the fruit does when both are exposed to the same water vapor partial pressure gradient. Because LP accelerates water vapor diffusion (Eq. (3.1); Figure 2.2), water loss from an insect life stage present on a commodity's surface at atmospheric pressure and 80% RH is likely to occur at the same rate in LP at 10 mmHg and 99.7% RH, and at 20 mmHg and 99.5% RH. The 0.3−0.5% RH gradient required for rapid loss of an insect's water in an LP chamber cannot be reliably measured with the most accurate commercially available humidity sensor (Section 5.1).

In LP studies with *Tenebriodes mauritanicus* (L.) larvae (Dumas et al., 1969) and when low [O_2] was tested on *Ephestia cautella* (Wlk.) pupae (Navarro and Donahaye, 1972), it was erroneously assumed that a humidity of 93–99% was sufficient to eliminate the likelihood of desiccation in an LP environment. Likewise Davenport et al. (2006), in tests performed in a "leaky" hypobaric storage chamber in which the humidity usually was no higher than 98%, assumed that a 98% RH was adequate to prevent insect desiccation. Nearly 98% of Caribbean fruit fly eggs and larvae maintained in this apparatus on nutrient agar media at 13°C were killed within 1 week, and 99.999% of their eggs were killed within 9.4 days at 15 mmHg and 10.6 days at 20 mmHg, whereas about 25% of the eggs survived for 14 days at 13°C, 760 mmHg, and 100% RH. At 13°C, the same rate of water loss should occur across a 1% RH gradient at 15 mmHg and a 50% RH gradient at atmospheric pressure (Eq. (3.1); Figure 2.2). When this experiment was repeated in a leak-free laboratory chamber at 15 mmHg, 99.7% RH, the fruit fly mortality was the same as previously reported (Davenport et al., 2006). This suggested that there might be a lethal factor associated with the agar medium, and when the experiment was repeated with fruit fly eggs present on plain agar or wet filter papers no measureable mortality occurred. Caribbean fruit fly eggs placed inside papaya, mango, or guava fruits, where they were protected from desiccation by the fruit's water-saturated intercellular atmosphere, displayed no mortality compared with controls (Davenport, 2007), but there was a small but significant mortality due to desiccation of eggs that were placed on the dry surfaces of the fruit. This study suggests that an extremely small reduction in the RH within an LP chamber may desiccate insect eggs oviposited on the surface of plant matter without causing an undue water loss to the stored commodity in the required time.

Insect-infested lettuce leaf samples were placed in vacuum chambers that were humidified with enclosed wetted towels, and after the chamber had been pumped down to 38 mmHg N_2 gas was injected into the chamber. Thirty minutes later the chamber was reevacuated to 38 mmHg and N_2 gas was again injected. This cycle was continuously repeated for the duration of the experiment. After 4 days at

5°C, mortality of *Nasonovia ribisnigri* (Mosley) and *Macrosiphum euphorbiae* (Thomas) aphid nymphs was 100%, leafminer (*Liriomyza langei* Frick) larvae mortality was >96%, and 60% of leafminer pupae had been killed (Liu, 2003). Liu suggested that because leafminer larvae feed internally in leaves they were not likely to be killed by dehydration, and therefore must have succumbed to the anaerobic condition. However, desiccation cannot be ruled out because each time the evacuated chamber was vented with (dry) N_2 the humidity within a lettuce leaf's intercellular system decreased, and when the chamber was reevacuated, expansion of the N_2 gas lowered the intercellular and chamber humidity. In a companion experiment performed at 10°C, the lettuce visually wilted during a 4-day treatment, indicating that the humidity was not saturated by enclosing wet towels.

A pressure of 15–20 mmHg at 2°C is useful for long-term lettuce storage, and within 52 h was 100% lethal to the green peach aphid, *Myzus persicae*, that infests this commodity (Aharoni et al., 1986). The aphid's mortality was pressure- and time dependent. It was concluded that death must be caused by low $[O_2]$ rather than desiccation or a reduction in pressure since the chamber was humidified. There was a significant difference in mortality at 20 and 50 mmHg within 24 h even though these treatments reduced the atmospheric pressure by nearly the same extent, 97.4% and 93.4%, respectively. The humidity was not measured and it was assumed that at both pressures the atmosphere was saturated and only the $[O_2]$ varied significantly, with 0.4% $[O_2]$ present at 20 mmHg and 1.23% $[O_2]$ at 50 mmHg. However, when the same chambers were used to store avocados and mangos in other experiments (Apelbaum et al., 1977b,c) their leak rate was high enough to increase commodity weight loss at pressures lower than 80 mmHg. This indicates that the humidity was not saturated during the lettuce experiment (Figure 2.1; Section 2.1). Therefore, it is possible that aphids on the surface of the lettuce were killed by desiccation, and this accounted for their more rapid mortality at 20 and 50 mmHg.

Repeated evacuations and venting during operation of the Metabolic System for Disinfection and Disinfestation (MSDD, Section 11.7) lowers the humidity in a vacuum chamber and kills various life stages of *Drosophila melanogaster* Meign, *Heliothis virescens* (F.), *Frankliniella occidentalis* Pergande, *Myzus persicae* Sulzer,

Tetranychus urticae Koch (two-spotted spider mite), and *Amblyseius cucumeris* Oudemans present on the commodity's surface or on agar in petri dishes (Lagunas-Solar et al., 2006). MSDD does not satisfactorily kill Mediterranean fruit fly (*Ceratitis capitata* Wiedemann), oriental fruit fly (*Bactrocera dorsalis* Hendel) or melon fly (*Bactrocera cucurbitae* Coquillett) eggs and larvae present inside papaya fruit where they are protected by the high intercellular humidity (Arévalo-Galarza and Follet, 2011). Arévalo-Galarza and Follet concluded that MDSS may be effective for surface-feeding insects, but not for insect life-forms that develop inside the host's tissue, even though hypoxia is more severe inside the tissue due to a horticultural commodity's respiratory oxygen consumption.

Desiccation is the best explanation for results obtained with the procedures and conditions used to eradicate museum pests (Selwitz and Maekawa, 1998). The Veloxy system for anoxic disinfestation, developed as part of the EU Project SAVE ART, consists of sealing items in envelopes made from a gas barrier plastic film, within which O_2 is reduced to <0.4% while the rest of the atmosphere is comprised of argon, helium, or nitrogen. The exposure time necessary to cause death to all life stages of the hardiest insects varies from 10 days using argon and helium to 21 days in nitrogen. Desiccation appears to be the main cause of insect death. Rapid insect dehydration occurs because spiracles open when the $[O_2]$ concentration is reduced to less than 1.0−1.5% or $[CO_2]$ is increased to >2%. Open spiracles enhance water loss up to 10-fold compared to the rate when spiracles are closed (Mellanby, 1934; Wigglesworth and Gillet, 1936; Bursell, 1957, Jay et al., 1971). At 20°C and 55% RH, the exposure time needed for the Veloxy system to kill insects is 3−4 days for adults, 7−10 days for eggs, and 14−20 days for larvae and pupa. Insects lose water much faster in the Veloxy system when argon or helium is used instead of nitrogen. It was suggested that this occurs because the molecular size of argon (1.22 Å) and helium (1.91 Å) is less than the size of nitrogen (2.31 Å). Enhanced water vapor diffusion (Figure 2.2) does not explain the improved efficacy of the Veloxy system using argon because according to Eq. (3.1) the binary diffusion coefficient of water vapor is similar in air, argon, and nitrogen and threefold higher in helium. According to the Matheson Gas Data Book, depending on the purity of the gas, compressed argon contains 1−5 ppm water, helium 0.1−5 ppm water, and nitrogen 0.2−3 ppm water.

11.2 GAS AND WATER VAPOR EXCHANGE THROUGH SPIRACLES AND TRACHEA

Air enters insect larvae, pupae, and adults through spiracles and diffuses to every part of the insect's body through interconnected, ramifying tracheal tubes, the smallest of which is approximately 2 μm in diameter. Thread-like ridges (taenidia) distributed spirally around the inner circumference of the trachea prevent them from collapsing when their interior pressure is reduced. At various places along their length, especially distally, the trachea are subdivided into capillaries (tracheoles), proximally approximately 1 μm in diameter, tapering to 0.1 μm or less at their ends. The tracheae and tracheoles are most abundant in areas of high metabolic activity where the tissue's O_2 need is greatest.

The insect body cavity contains the circulating blood (hemolymph) that is pumped by an open-heart system and bathes all organs and tissues. The main function of the hemolymph is to convey nutrient substances to the tissues and transfer waste products to the excretory organs. The blood does not contain a biochemical carrier of O_2 or CO_2 and therefore takes up no more of these gases than can be accounted for by aqueous physical solution. The hemolymph's part in the respiration of tracheate forms sometimes may be small but typically there are many cells separated from the nearest tracheal tubes by an appreciable distance and the blood acts as a carrier of O_2 to the deprived tissue by acquiring dissolved O_2 when it passes over tracheae.

Tracheoles are so small that the force of capillarity in their innermost ends is approximately 10 atmospheres (Eq. (10.1)). Therefore, an opposing force of that magnitude is required to prevent fluid from being withdrawn from adjacent tissues and transferred into the tracheal endings (Wigglesworth, 1930). The insect blood's osmotic pressure normally exceeds 10 atmospheres and prevents the capillary rise of water in the tracheoles, but lactate accumulates during periods of excessive energy consumption and causes water to be osmotically drawn into the tracheoles. When activity ceases or decreases, lactate is consumed, the fluid retreats, and air again fills the microcapillaries. Water intake limits the transfer of O_2 through tracheoles since gaseous diffusion occurs 10,000-fold more slowly through fluid than air (Weis-Fogh, 1964). The diameter of a plant tissue's intercellular spaces is much larger than 1 μm, so there is little tendency for capillary filling of plant intercellular spaces with water (Eq. (10.1)). Except when plant tissues become senescent fluid is prevented from entering the intercellular system by the cellular water potential.

Both spiracles and stomata close to protect against water loss and open to facilitate gas exchange, in insects to provide O_2 for respiration, and in plants to furnish CO_2 for photosynthesis. Desiccation closes both spiracles and stomata, and spiracles prevent dessication by opening for the shortest time needed to satisfy an insect's respiratory O_2 needs (Hazelhoff, 1927; Wigglesworth, 1935). Full spiracular opening in resting insects of various types and developmental stages is initiated by $1-20\%$ [CO_2] or less than $2-5\%$ [O_2] (Hazelhoff, 1927; Mellanby, 1934; Bursell, 1957; Burkett and Schneiderman, 1974), depending in part on the balance between the [O_2] and [CO_2] concentrations (Krafsur and Graham, 1970). These conditions lead to a relaxation of the opening muscle of "two-muscle" spiracles, and the ensuing closure tends to perpetuate itself because CO_2 accumulates and O_2 is consumed by respiration. Low [CO_2] opens and high [CO_2] closes plant stomata (Section 3.3), but more than $0.5-1\%$ [CO_2] opens stomata in darkness at atmospheric pressure (Wheeler et al., 1999; Levine et al., 2009; Section 3.3).

Drosophila melanogaster's tracheal system has been intensively investigated to provide information about the development of branched tubular networks used to transport gases and fluids in the mammalian lung, kidney, vascular system, and various glands. A fibroblast growth factor (FGF) and fibroblast growth factor receptor (FGFR) signaling pathway in both the insect tracheal system and mammalian lung patterns successive rounds of branching and outgrowth. The initial embryonic FGF patterning system is modified by genetic feedback controls and other signals at each stage of branching to give distinct branching outcomes (Metzger and Krasnow, 1999). *Drosophila's* tracheal system develops by sequential sprouting of primary, secondary, and terminal branches from an epithelial sac made up of approximately 80 cells in each body segment of the embryo. More than 30 genes have been identified and ordered into sequential steps controlling *Drosophila's* branching morphogenesis (Affolter and Shilo, 2000). The tracheal system forms by developing fine terminal branches at the beginning of the larval period (Jarecki et al., 1999). The FGF pathway then specifies the tracheal branching pattern by guiding tracheal cell migration during primary branch formation and activating later programs of finer branching at the ends of growing primary branches (Sutherland et al., 1996; Hacohen et al., 1998). *Drosophila's* branchless gene (*bnl*) is a key determinant of the tracheal branching pattern. It is expressed in clusters of cells surrounding the developing tracheal system at each position where a new branch will form and grow out. A homolog of mammalian FGFs is encoded by *bnl* and appears to function as a ligand for the breathless receptor, tyrosine kinase. The Branchless FGF gene induces secondary branching by activating the Breathless FGF receptor near the tips of growing primary branches.

Tracheal terminal branches have the capacity to sprout out projections toward [O_2]-starved areas (Ghabrial et al., 2003; Centanin et al., 2008). In insects, oxygen deprivation (hypoxia) induces the expression of genes that increases both the number of terminal tracheal branches (Wigglesworth, 1954; Jarecki et al., 1999) and the diameter of primary trachea (Locke, 1959; Loudon, 1989; Henry and Harrison, 2004). Hypoxic gene induction is mainly mediated by hypoxia-inducible factor (HIF), a heterodimeric α/β transcription factor composed of two basic-Helix-Loop-Helix-PAD (bHLH-PAS) subunits (Wang et al., 1995). HIF mediates cellular adaptations to low [O_2] in both mammals and *Drosophila* by means of prolyl-4-hydroxylase [O_2]-sensors that hydroxylate the HIF-alpha-subunit, promoting its degradation in normoxia. Mammals possess three HIF-prolyl-hydroxylases encoded by independent genes *phd1*, *phd2* and *phd3*. Expression of *phd1* is oxygen independent; *phd2* and *phd3* are induced by hypoxia and shut down HIF-dependent transcription upon reoxygenation. The *Drosophila* PHD locus, fatiga, encodes three HIF-prolyl-hydroxylase isoforms; FgaA is homologous to *phd2*, and FgaB and FgaC are similar to *phd3*. Hypoxia induces *fgaB* but not *fgaA* (Acevedo et al., 2010; Lavista-Llanos et al., 2002). Extra sprouting depends on the HIF-alpha homolog Sima and on the HIF-prolyl hydroxylase fatiga. Larval *Drosophila* cells experiencing hypoxia release *bnl* which induces spouting and extensive proliferation in nearby terminal tracheal branches and directs them to the hypoxic tissue (Jarecki et al., 1999).

Hypoxic induction of terminal tracheal branching in developing *Drosophila* embryos is maximally stimulated at atmospheric pressure by 3–5% [O_2] within 8 h. It decreases gradually as the oxygen concentration rises, and is inhibited at 1% [O_2], reflecting a general arrest of cell metabolism during severe hypoxia. Insect respiration ceases to be regulated below 1.6–3% [O_2], and in tissue-cultured *Drosophila* cells the hypoxic response is maximal at 0.1–1.0% [O_2] (Lavista-Llanos et al., 2002). Although insect life stages do not develop or grow during hypobaric storage at 10–20 mmHg (0.13–0.20% [O_2]), they sense the low-O_2 level and may undergo adaptations that help them to survive.

Nitric oxide (NO) and protein kinase G are involved in *Drosophila's* response to oxygen deprivation. By enhancing gaseous diffusion LP should decrease the amount of NO retained (Eq. (3.1); Figure 5.1), and this may interfere with an insect's response to hypoxia. Embryos and adult stages of *D. melanogaster* withstand and fully recover from hour-long exposures to hypoxic/anoxic challenges by sensing the falling partial pressure of O_2 within seconds via an NO signaling pathway (Wingrove and O'Farrell, 1999). Hypoxic exposure arrests the cell cycle and rapidly induces exploratory behavior in larvae. These responses are diminished by an inhibitor of NO synthase and by a polymorphism affecting a form of cGMP-dependent protein kinase. Conversely, these responses are induced by ectopic expression of NO synthase. Perturbing components of the NO/cGMP pathway alters both tracheal development and insect survival during prolonged hypoxia.

In larger more active insects, but not in terrestrial larvae and small winged insects, tracheal diffusion is supplemented by rhythmic muscular pumping of air through the tracheal system, stimulated by a scarcity of [O_2] or excessive [CO_2] (Krogh, 1941). The pulsation rate is highest during activity, is lower in pupae, decreases when the temperature is reduced, and almost ceases during hibernation. In many insects, the tracheae are expanded to form thin-walled air sacs that play an important role in ventilation. Gaseous exchange is enhanced by the alternating collapse and expansion of these air sacs due to rhythmic hemolymph pressure changes coupled with the synchronized opening and closing of spiracles.

Although ventilation often is continuous in insects, there may be extended periods during which all spiracles are closed, causing gas exchange and water loss through the spiracles to practically cease. Then, in inactive adults and some pupae, ventilatory movement of O_2

into and CO_2 from the trachea may occur by discontinuous gas exchange (DGS) in response to centers sensitive to CO_2 accumulation and to a lesser extent O_2 depletion. During DGS, the spiracular behavior has three phases: closed (C), flutter (F), and open (O) or burst (B) (Miller, 1973, 1981; Kestler, 1985; Mill, 1985; Lighton, 1994). Cyclic gas exchange (DGS) is characterized by a precise control of the spiracles, allowing the development of a negative pressure in the trachea with small amounts of air being periodically inspired. This pattern, known as passive suction ventilation, may be a mechanism for minimizing respiratory water loss from the trachea (Kestler, 1985; Jögar et al., 2004). In the cockroach (*Periplaneta americana*), water loss is greater during the burst phase releasing carbon dioxide than it is during the period of constriction flutter (Kestler, 1985; Machin et al., 1991). DGS probably does not occur at a low storage pressure because when O_2 is that scarce the spiracles open permanently to facilitate O_2 entry.

11.3 GAS EXCHANGE SYSTEMS OF INSECTS AND HORTICULTURAL COMMODITIES

The membranous coverings of insects and surfaces of leaves, flowers, and fruits have similar permeability properties and impede diffusive gas and vapor exchange to approximately the same extent. The insect's integument is comprised of a cuticle and single layer of epidermal cells covering the whole of the insect body and lining the tracheae. Adjacent epidermal cells are not tightly bound together and the lateral lymph spaces between them are liquid filled. Therefore, gas that exchanges through the insect integument must pass through a liquid phase just as it does entering through plant epidermal cells. The inner epicuticle of an insect is covered with an outer epicuticle comprised of polymerized lipid (cuticulin) over which is secreted an epicuticular wax layer having a composition similar to that in the plant cuticle. An insect's water loss occurs mainly through its respiratory surfaces (Mellanby, 1934; Wigglesworth and Gillett, 1936; Bursell, 1957). In both insects and plants, the waxy layers waterproof the surface, creating a transpirational resistance in the 20−400 s/cm range (Burg, 2004). The cuticular layer is practically impervious to O_2, but in insects it is sufficiently permeable to CO_2 to allow a considerable outward diffusion of this gas (Imms, 1949). The high CO_2 permeability and low O_2 permeability appears to be caused by the different water solubilities of these gases. The impermeability of insect cuticles to O_2 arises in the epicuticle, not from the wax layer that renders the cuticle impermeable to water (Buck, 1962). At atmospheric pressure, only a few percent of the total gaseous exchange in insects takes place through its cuticle, but the amount of water lost by this route is very large relative to that escaping through spiracles (Hadley, 1994; Lighton, 1994; Lighton and Berrigan, 1995), except possibly in pupal stages or in insects with an unusually high rate of water loss (Lighton et al., 1993; Williams et al., 1997).

Each insect trachea is lined with a cuticle and surrounded by an epithelial layer of cells that impede diffusive gas exchange to approximately the same extent that the epidermal cell layer and cuticle of plants hinder gas transport between the atmosphere and their intercellular system. In both insects (Herreid, 1980) and plants, the limiting factor in respiratory exchange is diffusion of O_2, which depends on oxygen's diffusion coefficient, the O_2-partial pressure difference between the atmosphere and mitochondria, and the system's total resistance to O_2 transport. Plant cytochrome oxidase and the terminal oxidase of insects have similar O_2 affinities (Chapman, 1998). Only a $2-5\%$ $[O_2]$ gradient is required between ambient air and the aeropyles and chorionic air spaces in insect eggs, open spiracles and tracheal endings in terrestrial larvae, pupae, and adult insects, and stomata or lenticles, and the center of a plant commodity's intercellular system to satisfy the aerobic respiratory $[O_2]$ needs without inducing fermentation in plant tissue or lactate production by insects (Krogh, 1920; Herreid, 1980; Burg, 2004). A high temperature exacerbates the effect of low $[O_2]$ because the metabolic demand for O_2 increases exponentially with a temperature rise (Gillooly et al., 2001), whereas the rate of oxygen diffusion into a tissue only increases by about 5% per 10°C increase (Denny, 1993).

There are two distinct phases in the transport of gases and vapors into and from insects, air tube diffusion through the tracheal system and tissue diffusion through liquid (Weis-Fogh, 1964). Even though the path for air tube diffusion is much longer, the resistance of an insect's liquid-filled pathway to diffusive O_2 entry exceeds that of the air-filled pathway because O_2 diffuses 100,000 times faster through air than it does through water or tissues (Weis-Fogh, 1964). A hypobaric pressure promotes O_2 and CO_2 diffusion to the same extent through an insect's air-filled trachea, spiracles, chorion layer, and aeropyles and through the stomata, lenticles, pedicel end-scar, intercellular spaces and barrier air layers of plant tissues. Nevertheless, insects are less suited than plant matter to satisfy the O_2 needs of their cells when O_2 is scarce at atmospheric pressure because gases diffuse much faster through air than they do through liquid, and in insects O_2 must exchange between tracheal air and the circulating blood in larvae, pupae, and adult stages, and through liquid in the egg yolk and cells of the amnion and serosa membranes. When the atmospheric pressure is reduced, a horticultural commodity's total O_2 mass-transport resistance approaches a lower limit established by cellular gas permeability because each plant cell communicates directly to the atmosphere through interconnected air-filled intercellular spaces, lenticles, stomata, and pedicel end-scars (Burg, 2004). A hypobaric condition decreases the mass-transfer resistance to a lesser extent in insects because O_2 must exchange between their tracheal air and a secondary fluid of considerable thickness, the circulating blood, in order to reach larval, pupal, and adult tissues that are not directly serviced by tracheoles. In addition, sufficient fluid may be present at the tracheolar endings, in the egg yolk, and in the egg's amnion and serosa membranes, to significantly hinder O_2 diffusion. Hypoxia and anoxia cause toxic levels of lactate and pyruvate to accumulate in insects (Price and Walter, 1987), but in plant commodities the relative quotient (RQ) remains close to unity and ethanol and acetaldehyde do not accumulate when as little as 0.08% $[O_2]$ is available during LP storage (Figure 8.1, right). Extensive air spaces are present in the chorion of an insect egg adjacent to the oocyte and this in part may explain why insect eggs are more tolerant than other developmental stages to low $[O_2]$ (Yoshida, 1975; Jay et al., 1971).

11.4 LETHAL EFFECT OF LOW $[O_2]$ AT ATMOSPHERIC PRESSURE

To reliably kill stored cereal insects and their larvae, the atmosphere must contain less than 0.9–5% O_2 (Oxley and Wickenden, 1963; Jay and Pearman, 1971; Shejbal et al., 1973; Banks and Annis, 1990; Fleurat-Lessard et al., 1990; Carpenter and Potter, 1994) with a balance of N_2, CO_2, and inert gases, or mixtures of these gases. These atmospheres function mainly by anoxic effect, and death presumably results from the accumulation of lactate and pyruvate produced by glycolysis (Price and Walter, 1987; Banks and Annis, 1990; Mitcham et al., 2006). An insect's survival may depend on its ability to accumulate and tolerate glycolytic products and reduce its metabolic rate (Carpenter and Potter, 1994; Mitcham et al., 2006). Most insect species and developmental stages display >95% mortality within <10 days in 0−1.0% [O_2] and 20−30°C (Abe and Kondoh, 1989; Banks and Annis, 1990). The time required for low [O_2] to cause insect lethality increases when the temperature is decreased and is longer for eggs and pupae than for larvae and adults (Yoshida, 1975; Jay et al., 1971).

CA has been used to kill insects (Thompson, 2010), but because the required low [O_2] and high temperature cannot be tolerated by fruits, vegetables, or cut flowers for very long, the most promising approach has been to expose the commodity to <1% [O_2] for short periods at a high temperature (Ke and Kader, 1989). The 0.2−0.4% [O_2] range used at atmospheric pressure to kill insects that infest mangos and papayas encompasses the [O_2] concentrations provided by the optimal pressure for long-term LP storage of these and most other types of plant matter.

11.5 LETHAL EFFECT OF HIGH [CO_2] AT ATMOSPHERIC PRESSURE

All life stages of most stored product insects, including eggs, are killed at 26°C by exposure to pure CO_2 for 10−48 h, to 60% CO_2 for 4 days, or 35% CO_2 for 10 days. Elevated [CO_2] sometimes enhances the action of low [O_2], but usually more than 35% [CO_2] is required for a significant effect (Fleurat-Lessard et al., 1990), and a concentration that high is injurious to almost all types of plant matter (Girsch, 1978; Calderon and Navarro, 1979; Burton, 1982). There usually is little or no influence on insect mortality from CO_2 or any other gas added to a <1% O_2 concentration (Banks and Annis, 1990).

11.6 LETHAL EFFECT OF ETHANOL AND OTHER VAPORS

The resistance of adult *D. melanogaster* to killing by ethanol vapor was first studied by Crozier et al. (1935). The time to death in air saturated with 59 mmHg ethanol vapor at 25°C varied from 27 min for 1-day-old adults to 10 min with 50-day-old adults. Heberlein (2000) found that ethanol can be added to the culture medium serving as *Drosophila's* food to ascertain whether a particular *Drosophila* strain was resistant to alcohol toxicity, and that a *Drosophila* population's resistance to alcohol can be rapidly increased by selectively breeding flies that survive exposure to high alcohol levels in their food. *Drosophila's* natural habitat is fermenting fruits containing high ethanol concentrations. The larvae consume yeasts growing on rotting fruits and have evolved resistance to fermentation products. *Drosophila* strains isolated from papaya are resistant to alcohol's toxic effects and can efficiently metabolize ethanol as an energy source (Greer et al., 1998). Ethanol reduces wasp oviposition into *Drosophila* larvae, and consumption of ethanol by fruit fly larvae kills wasp larvae growing in the larvae's hemocoel. Fly larvae seek out ethanol-containing food when they are wasp infected, indicating that they use ethanol as an anti-wasp medicine (Milan et al., 2012). When ethanol vapor is used to kill insects, a sufficient number of ethanol resistant mutants are likely to remain alive in the population to render the quarantine treatment ineffective.

Davenport (2007) saturated a sealed LP laboratory chamber with 59 mmHg of ethanol vapor at 25 mmHg, 13°C, and within 24 h killed 100% of Caribbean fruit fly (*Anastrepha suspensa*) eggs present on water-wetted filter paper. The ethanol vapor that condensed in the wet filter paper reached a 70% aqueous concentration, which is much higher than fruits tolerate. In addition to ethanol vapor (Section 11.7), volatile 3 to 12 carbon volatile aldehydes (Hammond and Kubo, 2001), acetaldehyde (Aharoni et al., 1979), and ethyl formate have been employed as pest control agents (Stewart and Aharoni, 1983).

11.7 METABOLIC STRESS DISINFECTION AND DISINFESTATION

Rapid evacuation and venting usually has no adverse effect on fruits, vegetables, cut flowers, or potted plants (Wheeler et al., 2011; Burg, 2004; Rennie et al., 2001), except that the compression that arises during rapid venting can squeeze juices from certain berries and alter the shape of grapefruits and other soft fruits. Releasing the pressure slowly, during 30 min, prevents these undesirable side effects (Burg, 2004). Calderon and Navarro (1968) proposed that insect death at a low pressure results primarily from an $[O_2]$ deficit and not from physical pressure on the insect. However, Lagunas-Solar et al. (2003, 2006) designed and tested an MSDD system that combines ethanol vapor with rapid cycles of low ambient pressure and high $[CO_2]$ to control pathogens and arthropod pests. He suggested that the response was due in part to anatomical changes in the insect caused by rapid pressure cycling.

The MSDD process is initiated with a physical phase consisting of a series of rapid decompression (evacuation) and compression (venting) cycles, during which pressure is first reduced and maintained at its lowest value for less than 1 min and then rapidly increased with CO_2 to a positive pressure for less than 1 min. This sequence is repeated for 10 cycles and lasts approximately 10–15 min, at which time environmental $[O_2]$ supposedly has been reduced to <0.0001% (v/v) and displaced from the arthropod's body.[1] The physical phase is followed by a chemical phase during which the chamber is evacuated to 74 mmHg and ethanol vapor is admitted for 3–4 h by an adiabatic expansion process, after which the chamber is flushed with a N_2/CO_2 gas mixture or filtered air.

Mortality of various life stages of *D. melanogaster* Meign, *H. virescens* (F.), *F. occidentalis* Pergande, *M. persicae* Sulzer, *T. urticae* Koch (two-spotted spider mite), and *A. cucumeris* Oudemans was high using the MSDD system (Lagunas-Solar et al., 2003, 2006). MDSS treatment with 125 mmHg and >99% CO_2 alone had no effect on mortality of second instar white peach scale (*Pseudaulacaspis pentagona* Targioni-Tozzetti) nymphs, but a combination of low pressure, high CO_2 and 75 mg/L of ethanol vapor was highly effective (Arévalo-Galarza et al., 2010). MSDD treatment of Hass avocados, including a 371-mg/L ethanol phase, was highly successful on the most resistant second/third instar larvae of surface-feeding long-tailed mealybug (*Pseudococcus longispinus*). Arévalo-Galarza and Follet (2011) tested the method on papayas infected with Mediterranean fruit fly (*C. capitata* Wiedemann), oriental fruit fly (*B. dorsalis* Hendel) and melon fly (*Bactrocera cucurbitae* Coquillett), separating the effect of the low pressure/ambient pressure cycles from the ethanol vapor phase treatment. Mortality was high when tephritid fruit fly larvae and adults were exposed to ethanol, and low when ethanol was withheld during the treatment, indicating that ethanol vapor can be highly lethal but fruit flies are quite tolerant of rapid pressure cycles and short periods

---

[1]This does not consider the escape of O_2 dissolved in the insect host. When mangos were stored at 15 mmHg, 13°C, the chamber $[O_2]$ exceeded the concentration in the air change for many days due to the slow release of $[O_2]$ initially dissolved in the mangos.

of low-pressure treatment alone. Papayas infested with fruit fly eggs or larvae and treated by MSDD produced fewer pupae than untreated control fruit, but a substantial number of individuals developed nonetheless. Arévalo-Galarza and Follet concluded that MSDD treatment has limited potential as a disinfestation treatment for internal-feeding quarantine pests such as fruit flies.

11.8 LETHAL EFFECT OF A LOW PRESSURE

The killing of various developmental stages of stored product insects by vacuum was first reported in 1925 by Back and Cotton, and the sensitivity of numerous types of insects to low pressures was listed at that time. Exposure of *Ephestia elutella* (Hbn.) and *Lasioderma serricorne* (F.) to high vacuum (Bare, 1948) resulted in mortality within the range cited in Back and Cotton (1925), and favorable results with high vacuum were also reported by El Nahal (1953) and Thornton and Sullivan (1964). Mortality of six species of common stored product insects at 18°C or 25°C and a low pressure was determined in containers filled with wheat (Calderon et al., 1966). The chambers were pumped down to 10 mmHg at 18°C, or 12 mmHg at 25°C, and within 120 h the low pressure caused 100% mortality of larval and adult forms of *Ephestia cautella* (Wlk.), *Oryzaephilus surinamensis* (L.), *Tribolium castaneum* (Hbst.), *Callosobruchus maculatus* (F.), *Sitophilus oryzae* (L), and *Trogoderma granarium* Everts. Adults of *E. cautella*, *O. surinamensis*, and *C. maculatus*, and adults and larvae of *T. castaneum* were killed within 7 h.

A nearly identical time is required to kill *T. castaneum* (Hbst.) adults at a low pressure and atmospheric pressure at the same O_2 concentration and temperature. Insect mortality was higher in LP at 25°C than it was at 18°C (Calderon et al., 1966), and the same temperature relationship arises when insects are exposed to low $[O_2]$ at atmospheric pressure and various temperatures (Yoshida, 1975). This presumably is due to the respiratory Q_{10}, which when the temperature is higher should cause a lethal dose of lactate to accumulate in the insect in a shorter time. A higher temperature also lowers the O_2 concentration available in the cellular liquid phase because it decreases oxygen's water solubility. These relationships cause efforts to kill insects by simultaneously lowering both the temperature and $[O_2]$ concentration to be offsetting.

When synchronized larval cultures of Caribbean fruit fly were incubated on agar culture media at 13.3°C and 1 atm pressure, the semi-gel media remained at pH 6.0 and turned brown due to the oxidation of phenolic materials and tyrosine by larval polyphenol oxidase. In LP, at 30 mmHg, lactic acid and pyruvic acid production by the fruit flies decreased the pH of the culture media to 3.5, and the agar remained clear because the O_2 partial pressure was too low to support polyphenol oxidase activity. When the larvae were removed after 14 days and incubated at room temperature, all 247 larvae had died during LP storage and none subsequently formed pupae, whereas at atmospheric pressure 80% of 379 larvae survived and formed pupae (Davenport and Burg, 1999, unpublished data). The experiment was repeated using synchronized egg cultures incubated at 13.3°C, either at atmospheric pressure or at pressures of 15 or 20 mmHg, flowing one 98% RH air change per hour. At atmospheric pressure, mortality gradually increased due to the low temperature, but 30% of the eggs survived during 15 days and were capable of hatching. Mortality reached 100% within 10–11 days at 15 mmHg and in 12–14 days at 20 mmHg (Davenport et al., 2006). These experiments were carried out in plastic vacuum chambers which had a high leak rate. Therefore, insect desiccation may have influenced the result. Pressures of 15 and 20 mmHg had no effect whatsoever on fly mortality when synchronized egg cultures of Caribbean fruit flies were incubated inside papaya or mango fruits kept at 13°C in a leak-free apparatus (Davenport, 2007). Phillips et al. (2002) studied the mortality of apple maggot eggs (*Rhagoletis pomonella*) when vacuum jars containing infested fruits were evacuated to 25 mmHg and the pressure rose to 40 mmHg during 48 h (leak rate = 0.3 mmHg/h). The RH was not controlled or measured. In 48 h, approximately 95% of the maggot eggs died.

11.9 IONIZING RADIATION (ALSO SEE SECTION 12.8)

Ionizing radiation from high-energy electron beams, gamma rays from cobalt 60, and X-rays disrupt an insect's normal cellular function by breaking chemical bonds within DNA and other molecules. The International Commission on Microbiological Specifications for Foods (ICMSF) established that X-rays and ionizing radiation from cobalt 60 penetrate approximately 20 cm into fresh plant products and boxes (Frazier and Westhoff, 1988), making this an effective way to provide quarantine security by preventing insect reproduction rather than by killing the insect pest immediately. Electron beam irradiation has a much lower penetration power than gamma radiation, but because the electron beam linear accelerator concentrates and accelerates electrons to 99% the speed of light at 5–10 million electron volts (MeV), only a few second exposure is required. Electron beams have the advantage that they are produced electrically, do not require a radioactive source, and have a utilization efficiency of 40–60% vs. 10–25% for gamma rays. X-rays are inefficient and have the highest cost. The US Food and Drug Administration (FDA) approved radiation doses of up to 1 kilogray (kGy)[2] for the preservation and disinfestation of fresh fruits and vegetables in 1986, and in 2006 USDA-APHIS published a pioneering rule permitting generic low-dose radiation quarantine treatments for controlling insects in all fresh horticultural commodities (Follet, 2010). A generic radiation treatment is a single treatment that controls a broad group of pests without adversely affecting commodity quality. The 2006 rule approved radiation doses of 150 Gy for any tephritid fruit fly, and 400 Gy for all other insects except the pupa and adult stages of Lepidoptera (moths and butterflies).

Irradiation of Haden mangos at 100 Gy provides a high level of quarantine security against Mediterranean fruit fly, and it may also suffice for avocados (Torres-Rivera and Hallman, 2007). A dose of 150 Gy applied to melon fly third instars in papayas, and 100 and 125 Gy to Mediterranean fruit fly and oriental fruit fly late third instars, respectively, resulted in no survival of the adult stage (Follet and Armstrong, 2004). APHIS published a final rule in 2007 to allow the importation of commercial shipments of fresh irradiated mangos from India into the continental United States. The mangos are first inspected for the presence of pests by an APHIS preclearance officer in India, and if no pests are found, the fruit is authorized to be treated with a specified dose of irradiation at an APHIS-certified facility prior to export. If live pests are found during the preclearance inspection, the shipment is refused treatment and cannot be exported to the United States. The mangos must be packed in insect-proof boxes, safe-guarded after radiation treatment to prevent reinfestation, and accompanied by a phytosanitary certificate issued by the National Plant Protection Organization (NPPO) of India certifying that the treatment

---

[2]The gray (Gy) is defined as the absorption of 1 J of ionizing radiation by a 1 kg weight (1 J/kg). A kilogray (kGy) is an absorbed ionizing radiation dose equal to one thousand gray (1000 Gy). One rad (radiation absorbed dose) = 1/100th Gy.

and inspection of the mangos were made in accordance with the regulations. A radiation facility plant in Iowa (USA), which has been authorized to irradiate imported Pakistani mangos, charges importers by the hour rather than by the weight of fruit, and the hourly pricing may make mangos imported from Pakistan prohibitively expensive (Sarwar, 2012). Aphis has approved building an electron beam radiation facility in the McAllen Free Trade Zone, Texas, on the Mexican border.

Irradiation of mature preclimacteric Alphonso mangos in N_2 or air with 25 krads extends the fruit's storage life by 6 days (Dharkar et al., 1966). Banana ripening is delayed by $20-40$ kGy without affecting quality, and the fruits can be ripened later with ethylene (Thomas and Moy, 1986; Kao, 1971; Kader, 1986). Papaya ripening is delayed by 0.75 kGy, the fruit's green color is modified to a lighter more intense tone, and yellow color development is improved without affecting weight loss, pH, and total solids (Pimentel and Walder, 2004). Similar results have been obtained with guava and other subtropical and tropical fruits (Akamine and Moy, 1983; Thomas and Moy, 1986. The shelf-life of cv. Matsutake mushrooms is extended by γ-irradiation (Aoki et al., 1974). More than 1 kGy inhibits the ripening of temperate-zone fruits such as apple, pear, and apricot, stimulates the respiration of climacteric pears, peaches, and nonclimacteric cherries and strawberries, and causes uneven ripening and excessive softening. More than 4 kGy stimulates and less than 4 kGy inhibits ethylene production by fruits. Higher doses reduce a fruit's sensitivity to the ripening action of ethylene (Kader, 1986), and >1 kGy induces physiological disorders in bananas, avocados, grapes, olives, Kadota figs, cucumbers, summer squash, peppers, artichokes, lettuce, endive, and sweet corn (Kader, 1986; Bramlage and Couey, 1965; Bramlage and Lipton, 1965; Lipton et al., 1967; Maxie and Kader, 1966).

E-beam irradiation of Tommy Atkins mangos at 3.0 kGy caused no detriment to overall sensory and chemical quality, but during a subsequent 21-day storage the fruit's ascorbic acid content declined by $50-70\%$ and the level of phenolics increased by $18.3-27.4\%$ (Moreno et al., 2007). In another study, 2 kGy γ-irradiation caused insignificant changes in ascorbic acid, titratable acidity and total sugar in Hortis Gold and Pepinos papayas, and Kent and Zill mangos (Thomas and Beyers, 1978). The production of terpenes by Chok Anan mango fruits was reduced after irradiation with 0.3 kGy followed by 10 days storage at 13°C (Laohakunjit et al., 2005). The FDA concluded that foods irradiated at dose levels up to 1 kGy are safe for human consumption, and that generally there is no effect on the nutrients of foods irradiated with up to 10 kGy of γ-irradiation. Free radicals are formed when food is irradiated, but in wet foods they disappear within a fraction of a second (Kilonzo-Nthenge, 2012).

CHAPTER 12

Fungi and Bacteria

Erwinia atroseptica, E. carotovora, and 25 fungal species are responsible for the major postharvest diseases of plant commodities (Eckert and Ratnayake, 1983).

12.1 EFFECT OF LOW [O₂] AT ATMOSPHERIC PRESSURE

Less than 1% [O_2] is required to substantially inhibit the growth of obligate aerobic and microaerophilic bacteria. Growth of *Escherichia coli* 0157:H7 on raw salad vegetables is not affected by 3% [O_2] (Abdul Raouf et al., 1993) and development of *Pseudomonas fluorescens, E. carotovora,* and *E. atroseptica* on buffered asparagine–yeast extract broth is reduced in 1–3% [O_2] and decreases linearly below 1% [O_2] (Figure 12.2).

Tabak and Cooke (1968) reviewed literature from 1889 through 1968 describing O_2 and CO_2 effects on the growth, sporulation, and spore germination of fungi and yeasts at atmospheric pressure. *Ascophanus carneus, Phoma, Rhizoctonia solani, Ophiobolus graminis, Claviceps purpurea, Neurospora sitophila, Penicillium roqueforti, P. expansum, Geotrichum candidum, Aspergillus flavus,* and *A. niger* were listed as fungi whose growth is inhibited in low [O_2]. Air's O_2 content must be reduced to <1% and often to <0.2% before growth of most fungi is strongly inhibited (Follstad, 1966; Banks and Annis, 1990; Cochrane, 1958; Imoolehin and Grogan, 1980; Figure 12.1). Some fungi grow as well in 0.5–5% [O_2] as in air, and among 18 species of fungal pathogens tested at 5.5–12.5°C, growth of the majority was not inhibited by 2.3% [O_2] (El-Goorani and Sommer, 1979). Atmospheres containing 3% [O_2], and even less, do not inhibit the growth of *Rhizopus* and *Alternaria* cultures (Parsons et al., 1970). Growth of *Rhizopus, Penicillium, Phomopsis,* and *Sclerotinia* cultures is inhibited by total anaerobiosis, but these fungi develop at a nearly normal rate in 99% [N_2] + 1% [O_2] (Ryall, 1963). The colony diameter of *Botrytis cinerea, Alternaria tenuis, Cladosporium herbarum,* and *Rhizopus stolonifer* growing on solid media is decreased by one-third to one-half

Hypobaric Storage in Food Industry. DOI: http://dx.doi.org/10.1016/B978-0-12-419962-0.00012-7

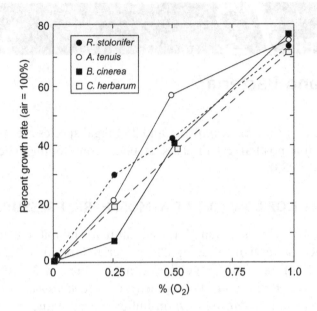

Figure 12.1 Average growth response of R. Stolonifer, A. tenuis, B. cinerea, and C. herbarum cultured on agar medium at 15°C in atmospheres containing 1% [O₂] or less, at atmospheric pressure. Original (in public domain) reproduced from Follstad (1966); Figure 1.1.

in 0.5% [O$_2$], and by even more in 0.25% [O$_2$] (Follstad, 1966; Figure 12.1), and the development of *Botrytis* on carnations and straw-berries is inhibited to this same extent at a comparable low O$_2$ partial pressure (Burg, 2004). Growth of *Gloeosporium album* and *Fusarium oxysporum* in liquid media is slowed when [O$_2$] is lowered to 2.5% (Lockhardt, 1969). The growth of *A. tenuis, B. cinerea, C. herbarum, Fusarium roseum,* and *R. stolonifer* that developed in liquid gluco-se—salt media was reduced to zero when [O$_2$] in the aerating mixture was decreased to between 0 and 4% in the absence of CO$_2$ (Wells and Uota, 1970). Cultures of *Phytophthora cactorum* flushed with N$_2$ gas containing 0.2% residual O$_2$ developed at 30% of the normal rate in air (Covey, 1970). Mold development usually is reduced substantially in the 0.3% [O$_2$] concentration present in technical grade N$_2$ gas, but to completely inhibit fungal growth it often is necessary to provide even lower [O$_2$] (Paster, 1990). An anaerobic condition prevented spore germination of *Gloeosporium musarum* (= *Colletotrichum musae*), but only slightly inhibited its vegetative growth (Goos and Tschirsch, 1962). *A. tenuis, B. cinerea,* and *Colletotrichum coccodes* inoculated on tomatoes grew at the same rate in 3% [O$_2$] or air, but in 0.25% [O$_2$] growth was significantly slowed (Parsons and Spalding, 1971).

Anoxia prevents or inhibits mold growth, but does not kill fungal spores, and vegetative cells are able to renew their growth when oxygen is reintroduced. Obligate aerobic bacteria only grow in the presence of O_2 and cannot carry out fermentation, but they are not killed by a lack of O_2. Obligate anaerobic bacteria do not carry out oxidative phosphorylation and are killed by O_2 because they lack enzymes such as catalase, peroxidase, and superoxide dismutase that detoxify peroxide and oxygen free radicals produced during metabolism in the presence of O_2. Aero-tolerant anaerobes are bacteria that "respire" anaerobically, but can survive in the presence of O_2. Facultative anaerobic bacteria can perform both fermentation and aerobic respiration, and in the presence of O_2 their anaerobic respiration ceases and they respire aerobically. Microaerophilic bacteria grow well in low $[O_2]$ concentrations, but are killed by higher concentrations (Murray, 2011).

CA and MA usually cannot prevent decay by lowering $[O_2]$ because at atmospheric pressure $<1-2\%$ $[O_2]$ at atmospheric pressure causes low-$[O_2]$ injury to most types of plant matter (Burton, 1982; Table 8.1).

12.2 EFFECT OF $[CO_2]$ AT ATMOSPHERIC PRESSURE

The sporulation and mycelial growth of many fungi that decay plant matter is inhibited by $20-50\%$ $[CO_2]$ (Tabak and Cooke, 1968; Ulrich, 1975; Tian et al., 2001; Couey et al. 1966). More than 35% $[CO_2]$ often is required (Paster, 1990), and fungal growth inhibition seldom reaches 50% at the CO_2 concentrations tolerated by most and types of plant matter. At $12.8°C$, tomatoes (cv. Homestead) inoculated with *G. candidum*, *A. tenuis*, *B. cinerea*, or *C. coccodes* decayed less in air than in 3% $[O_2]$, and adding 5% $[CO_2]$ to the CA atmosphere did not decrease the total number of decayed fruits, but in 0.25% $[O_2] + 5\%$ $[CO_2]$ decay was reduced (Parsons and Spalding, 1971).

Some fungi require CO_2 to develop and sporulate (Tabak and Cooke, 1968). *Aspergillus oryza's* growth is not inhibited by 10% $[CO_2]$, *Mucor* and *Aspergillus* require CO_2 to grow, and *A. niger* spores do not germinate in the absence of CO_2. Low $[CO_2]$ promotes and much higher concentrations inhibit *Penicillium* growth. Development of *Verticillium albo-atrum* is severely curtailed in a CO_2-free atmosphere (Hartman et al., 1972). *Fusarium roqueforti* tolerates 75% $[CO_2]$, its growth is promoted by 3−10% [CO_2], and germination of its spores is stimulated by 4−16% [CO_2] (Stover and Frieberg, 1958; Toler et al., 1966). *A. tenuis, B. cinerea, C. herbarum*, and *R. stolonifer* develop reasonably well in 10−20% [CO_2]. *Botryodiploidia theobromae*, a causative agent of banana stem-end rot and anthracnose, is suppressed by 2.3% [O_2] to the same extent in the presence or absence of 5% [CO_2] (El-Goorani and Sommer, 1979). Less than 0.03% [CO_2] promotes and relatively high [CO_2] inhibits *F. roseum's* mycelial growth and spore germination (Wells and Uota, 1970). Even 100% [CO_2] does not inhibit the growth and reproduction of yeasts. A direct growth inhibition of disease organisms by MA packaging or with CO_2 added during CA storage has only been demonstrable with strawberries, cherries, blueberries, and a few other horticultural commodities able to withstand more than 10−20% [CO_2] without developing off-flavors, unpleasant odors, and aberrant colors (Wells, 1970; Couey and Wells, 1970; Prince, 1989).

Growth of aerobic and facultative anaerobic bacteria such as *P. fluorescens, E. atroseptica*, and *E. carotovora* is inhibited by 10−100% [CO_2] (Figure 12.2), but no amount of CO_2 suppresses the growth of anaerobic or facultative anaerobic *Streptococci* and microaerophilic *Lactobacilli* (Johnson, 1974; Gill and Harrison, 1989). Development of many aerobic bacteria and molds is stimulated by low [CO_2] and retarded or prevented by oxygenation with CO_2-free air (Rochwell and Highberger, 1927; Valley and Rettger, 1927; Valley, 1927; Rahn, 1941; Krebs, 1943; Tabak and Cooke, 1968). Anaerobic bacterial growth may also require CO_2, especially among coliforms (Thimann, 1955). Six isolates of *E. atroseptica*, a facultative anaerobe, could not grow in CO_2-free air at 21°C, and trace levels of growth occurred in only one of six isolates of *E. carotovora* (Figure 12.2; Wells, 1974). Growth of *P. fluorescens*, an aerobic bacterium, doubled when 3% [CO_2] was provided to cultures incubated in 1% [O_2] (Wells, 1974). The generation time of *Streptococcus haemolyticus* (Pappenheimer and Hottie, 1940) is increased by low [CO_2], and the generation time of *Pseudomonas aeruginosa* is decreased 2- to 3-fold in gas mixtures containing high [CO_2] (King, 1966—referred to in Wells, 1974). The mean generation time of *Aerobacter aerogenes* was 320 min in CO_2-free air, 100 min in 0.04% CO_2, and 50 min in 0.16% CO_2 (Monod, 1942). CO_2 incubators are used in bacteriology laboratories to *accelerate* the growth of CO_2-dependent bacteria.

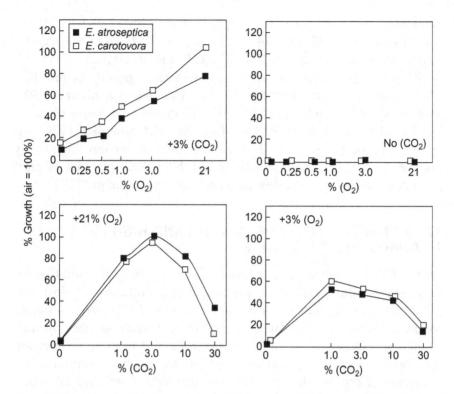

Figure 12.2 Growth on asparagine yeast extract broth of E. carotovora and E. atroseptica at 21°C during 16 and 24 hours, respectively, at atmospheric pressure (Wells, 1974). Original Figure 1.1AD (in public domain) reproduced from Wells, 1974.

12.3 COMBINED EFFECT OF LOW $[O_2]$ + HIGH $[CO_2]$ AT ATMOSPHERIC PRESSURE

Low $[O_2]$ and high $[CO_2]$ do not usually cause an additive inhibition of fungal growth and spore germination. Instead, these conditions tend to be mutually offsetting, and any benefit low $[O_2]$ provides for decay control is likely to be counteracted by a tolerable $[CO_2]$ concentration (Banks and Annis, 1990). The growth of G. album and F. oxysporum in 2.5–15% $[O_2]$ is stimulated by 5–10% $[CO_2]$ (Lockhardt, 1967, 1969). Anaerobic growth of G. candidum occurs if CO_2 is added, but not in its absence, and at all $[O_2]$ concentrations 3% $[CO_2]$ stimulates this pathogen's development by 33% (Wells and Spalding, 1975). Germination of Alternaria alternata, B. cinerea, C. herbarum, and F. roseum spores is stimulated by 4% $[CO_2]$ (Wells and Uota, 1970), the development of these fungi in 1–2% $[O_2]$ is promoted by 4–16% $[CO_2]$, and F. oxysporum's growth is reduced by decreasing $[O_2]$ and stimulated if low $[O_2]$ is supplemented with 2.5–15% $[CO_2]$ (Lockhardt, 1968). Growth of several species of *Phytophthora* in 1% $[O_2]$ is promoted by 5% $[CO_2]$ (Mitchell and Zentmyer, 1971) and 3–5% $[CO_2]$ promotes *Rhizopus* and *Alternaria* growth in low $[O_2]$ (Parson et al., 1970). Proliferation of *Monilinia fructicola* is 50% suppressed in 2.3% $[O_2]$, and this effect is almost completely obviated by the addition of 5% $[CO_2]$ (El-Goorani and Sommer, 1979). At atmospheric pressure, CA and MA are ineffective in controlling decay because of these O_2/CO_2 interactions and the limited tolerance of most plant commodities to ultra-low $[O_2]$ and moderate or high $[CO_2]$.

12.4 INDIRECT EFFECTS OF LOW $[O_2]$ AND HIGH $[CO_2]$ AT ATMOSPHERIC PRESSURE

Burton (1982) claimed "no CA atmosphere in which commodities can be stored without injury gives any appreciable protection against fungal attack, and the same can be said of rotting by bacteria, different species of which cover the whole range from obligate aerobes through facultative anaerobes to obligate anaerobes." Cherries, blueberries, strawberries, and a few other commodities are notable exceptions, and secondary effects on disease resistance attributable to delayed senescence and ripening often have been cited as "evidence that CA prevents decay" (Thompson, 2010). Storing mature green tomatoes in 3% $[O_2]$ + 3% $[CO_2]$ to delay ripening markedly reduces decay caused by *Rhizopus* and *Alternaria* during 6 weeks at 13°C, even though cultures of these molds are hardly affected by 3% $[O_2]$, and addition of 3% $[CO_2]$ promotes their growth (Follstad, 1966; Wells and Uota, 1970).

12.5 EFFECT OF LOW $[O_2]$ + LOW $[CO_2]$ DURING HYPOBARIC STORAGE

Growth and spore germination of *Penicillium digitatum*, *P. expansum*, *P. spinulosum*, *P. diversum*, *B. cinerea*, *Trichothecium roseum*, *R. cinerea*, *A. alternata*, and *G. candidum* var. *citri-aurantii* cultures was only slightly inhibited at 100 mmHg, 23°C (2.2% $[O_2]$). Growth and spore germination progressively decreased below 100 mmHg and ceased at 25 mmHg (0.14% $[O_2]$) in *B. cinerea*, *G. candidum*, and *A. alternata* (Figure 12.3; Wu and Salunkhe, 1972a; Salunkhe and Wu, 1975; Apelbaum and Barkai-Golan, 1977; Borecka and Olak, 1978). Development of *Colletotrichum gloeosporioides* cultures was severely

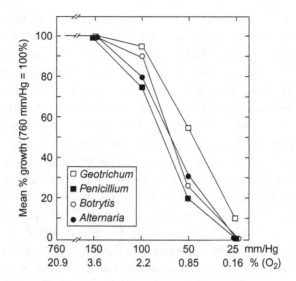

Figure 12.3 Effect of hypobaric storage on colony growth of G. candidum var. citri-aurentii, P. digitatum, B, cinerea, and A. alternata at 23°C. [O₂] is computed assuming 95% RH. From original Figure 1.2 © The American Phytopathological Society, Apelbaum and Barkai-Golan (1977). Reproduced by permission.

inhibited and sporulation prevented during 18 days at 10°C, 15 mmHg (0.14% [O₂]). A fibrillar network that covers this mold's mycelium at atmospheric pressure was absent (Chau and Alvarez, 1983).

Both in low [O₂] at atmospheric pressure and the same oxygen concentration at low pressure, there is a close correlation between a microbe's ability to grow on agar or in liquid media, and the decay the orgasm causes in the host commodity stored at the same condition. Growth and spore germination in *B. cinerea* cultures and development of *B. cinerea* in carnation flowers (Uota and Garzsi, 1967) or strawberries (Couey et al., 1966) almost ceases at 0°C in 0.1−0.25% [O₂] both at atmospheric pressure and in 0.15−0.29% [O₂] at 10−15 mmHg (Jamieson, 1980a; Follstad, 1966). *C. gloeosporioides*, the causative agent of papaya anthracnose, develops normally cultured at 10°C and 150 mmHg (3.9% [O₂]), but at 10°C and 15 mmHg (0.14% [O₂]) this organism's growth and sporulation are severely inhibited (Chau and Alvarez, 1983; Chau, 1981). During a 21-day papaya storage at 10°C and atmospheric pressure, 30−50% of conidia germinated and formed appressoria, and a mycelial mat with hyphae covered the fruit's surface (Chau and Alvarez, 1983). When mature green papaya fruits inoculated with *C. gloeosporioides* were stored for 21 days at 10°C, 15 mmHg (0.16% [O₂]), pathogen development was suppressed for 5 days after the fruits were transferred to atmospheric air and ripened at room temperature, and fewer than 5% of conidia germinated and formed appressoria within 7 days at 10°C. Infection was not observed until 2 weeks after these fruits were removed from LP. In a comparison of 10°C shipments of papayas from Hawaii to New York City in conventional refrigerated intermodal containers versus Grumman/Dormavac containers at 20 mmHg (0.3% [O_2]), a low pressure reduced the development of papaya stem-end rot and anthracnose caused by *C. gloeosporioides* by 63%, the incidence of stem-end rots due to *Mycosphaerella* a*scochyta caricae-papayae* and *B. theobromae* by 55% and *C. gloeosporioides* peduncle infections by 45% (Alvarez, 1980). Bacterial soft rot developed in asparagus at 0°C and pressures ranging from 760 to 80 mmHg (21−2.1% [O_2]), but not in LP at 20−40 mmHg (0.3−0.85% [O_2]). This indicates that insufficient [CO_2] and [O_2] was present at 20−40 mmHg to support the growth of *E. carotovora*, the causative agent of bacterial soft rot in asparagus (Dilley, 1977; Figure 12.2). Direct suppression of mold and bacterial growth during LP storage has also been demonstrated in carambola, cucumber, mango, strawberry, asparagus, papaya, cherry, green beans, and cut flowers (Burg, 2004).

Schuerger and Nicholson (2006) and Schuerger et al. (2013) measured the growth of 26 strains of 22 bacterial species cultured on agar media in petri dishes present in a vacuum bell jar. The pressure was controlled ± 0.5 mbar (± 0.375 mmHg) by turning the vacuum pump on/off in response to high/low-pressure limit sensors. The lower limit of pressure that could be reached with the bell jar present in a 30°C incubator was 25 mbar (18.7 mmHg), and at that pressure the agar visibly desiccated during a 48-h test. The saturated vapor pressure of water is 25 mbar at 21°C (Table 9.1, lower). Therefore, vacuum cooling had to lower the temperature of the agar media in the petri dishes to 21°C before the pressure could be decreased to 25 mbar. This caused the agar to lose 1.3% of its water. The air temperature in the bell jar should have equilibrated mid-way between 30°C and 21°C, at approximately 25.5°C, and the maximum water vapor pressure this air could have contained was the saturated vapor pressure of water at 21°C. Therefore, the air's relative humidity (RH) was approximately 76%. Heat was transferred from the warmer bell jar to the cooler agar plates by radiation and convection (Sections 2.1 and 2.2), establishing a thermodynamic equilibrium when the heat transfer rate to the agar plates equalled the rate at which heat was removed from the plates by evaporative cooling caused by water continuously boiling from the agar's surface. In 48 h, heat transferred from the bell jar by radiation across a 9°C temperature difference, evaporated slightly more than 7% of the agar's water (Eq. (4.9); Section 4.2). Convection from the chamber air to the agar caused an additional 1% water loss in 48 h (Burg, 2004). The Q_{10} for bacterial growth varies between 1.9 and 3.4 depending on the bacterial species and conditions. Therefore, a 9°C temperature decrease should reduce growth by 47−71%. Instead, growth of all bacterial cultures was totally prevented or drastically inhibited at 25 mbar, whereas it was only slightly inhibited at and above 50 mbar. The large inhibition of bacterial growth at a pressure/temperature combination that caused water to boil, and a much smaller inhibition at a slightly higher pressure that was below water's boiling point at 21°C, indicates that a decrease in water activity (a_w) was the major cause of the growth inhibition at 25 mbar. Water activity equals p_{cell}/p_{air}, where p_{cell} and p_{air} are the vapor pressure of water in the bacterial cells and bell jar air, respectively. Since under the conditions of this study $p_{air} = RH/100 \approx 0.76$, therefore $a_w \leq 0.76$. Most gram-positive and gram-negative bacteria cannot grow at $a_w \leq 0.90-0.91$ and $a_w \leq 0.95-0.97$, respectively (Troller et al., 1984). The bacteria were able to grow without water stress at ≥ 50 mbar because the air temperature adjacent to the agar surface would be close to 29.5°C and the RH of the chamber air is 96.5% ($a_w \approx 0.965$). Bacteria survive water activity stress by osmoregulation, gene transcription regulation, protein synthesis, accumulation of osmo-protectants, such as proline or glycine betaine, and by altering their cell membrane transport properties (Csonka, 1989). Essentially all of the bacteria survived and were able to resume near-normal growth when the agar plates were returned to atmospheric pressure. During 24 h in liquid culture, growth of *Bacillus subtilis* and *E. coli* inside a polycarbonate vacuum desiccator present in a 30°C incubator was progressively inhibited when the pressure was lowered below 100 mbar. At 25 mbar, the inhibition was 85% with *B. subtilis* and 30% with *E. coli*. The bacteria were grown in 10 mL of aqueous media contained in 18 mm diameter screw-top tubes. *E. coli's* Q_{10} for growth is 2.0 between 20°C and 30°C (Ng, 1969), and this could explain its entire growth inhibition after the tubes cooled to 21°C at 25 mbar, *B. subtilis* would need to have a Q_{10} of 3.4 to account for its growth inhibition at 25 mbar. The water volume in each tube did not decrease during 24 h at 50, 75, and 1013 mbar, but 10−15%% of the water boiled and escaped through the loosened tube caps at 25 mbar. In this liquid system, $a_w = p_{cell}/p_o$, where

p_{cell} is the vapor pressure of water in the bacterial cells, and p_o the vapor pressure of the liquid media's water. If water was lost from the 25-mbar tubes at a constant rate, *E. coli* would have reached its minimum a_w for growth in 8–12 h, and *B. subtilis* in 16–24 h. The average amount of liquid water boiling from a *single* tube at 25 mbar increased the pressure in the polycarbonate desiccator by 6.7 mbar/h,[1] and air leakage into this type of desiccator increased the pressure by at least 2.7 mbar/h. These effects, O_2 consumption by the bacteria, and intermittent evacuation to keep the pressure at 25 mbar influenced growth by creating a nearly anaerobic condition.

A low $[CO_2]$ condition arises during hypobaric storage because LP facilitates respiratory CO_2 diffusion from plant matter (Figure 2.2; Eq. (3.1)) through the intercellular system, lenticles, pedicle-end scars, and via stomata when LP causes them to open in darkness (Section 3.3). Simultaneously, the LP chamber is continuously ventilated with rarified air changes in which the atmospheric CO_2 content has decreased due to expansion during entry into the storage chamber. Apelbaum and Barkai-Golan (1977) suggested that the resultant ultra-low CO_2 concentration might augment the antifungal effect caused by a simultaneous reduction in $[O_2]$. They found that the growth inhibition of *P. digitatum*, *B. cinerea*, *A. alternata*, *Diplodia natalensis*, and *G. candidum* var. *citri-auranti* cultures in 0.8% $[O_2]$ was greater at 50 mmHg than at atmospheric pressure (Table 12.1), and that

Table 12.1 Effect of LP on Postharvest Pathogen Growth at Equivalent $[O_2]$ Partial Pressures and Different Total Pressures

Fungus	% Growth Relative to Growth at 760 mmHg (21% $[O_2]$)	
	50 mmHg (0.8% $[O_2]$)	760 mmHg (0.8% $[O_2]$)
Penicillium digitatum	35	55
Botrytis cinerea	37	60
Alternaria alternata	46	82
Diplodia natalensis	53	83
Geotrichum candidum var. *citri-aurantii*	75	100

Cultures were incubated at 21.1°C for 8 days on potato dextrose agar, either ventilated with water-saturated O_2 at 50 mmHg (pO₂ = 0.008 atm), water-saturated air (pO₂ = 0.21 atm) at atmospheric pressure (760 mmHg), or water-saturated air in which the pO₂ was lowered to 0.008 atm.
Reproduced by permission, from the original Table 1.1 of Apelbaum and Barkai-Golan, 1977) (© The American Phytopathological Society).

[1]The publications do not indicate how many tubes were present in the desiccator.

G. candidum did not proliferate at 23°C and a pressure of 25 mmHg (0.14% [O_2]) even though this mold's growth is stimulated at atmospheric pressure and that same [O_2] tension (Lockhardt, 1967; El-Goorani and Sommer, 1979). Likewise, Wu and Salunkhe (1972a) reported that the inhibition of growth and sporulation in *Penicillium expansum*, *Rhizopus nigricans*, *A. niger*, *Botrytis alli*, and *Alternaria* caused by 2.3% [O_2] is smaller at 102 mmHg than it is at atmospheric pressure.

Lougheed et al. (1977, 1978) claimed that few commodities could tolerate the degree of hypoxia required to prevent microbial growth during hypobaric storage, that LP lacks the benefit adding CO_2 provides for decay control in CA, and the nearly saturated RH required for LP storage favors fungal and bacterial growth. To the contrary, many studies have shown that CO_2 removal and a high noncondensing RH reduce decay (van den Berg and Lentz, 1978), and although the low O_2 partial pressures required to prevent decay during CA storage typically are injurious to plant matter, they do not cause low-[O_2] damage in LP (Table 8.1).

12.6 A MODEST PRESSURE CHANGE ACTIVATES ENZYMES THAT SUPPRESS MOLD GROWTH

Romanazzi et al. (2001, 2003) reported that a 1- to 24-h hypobaric treatment at 20°C and 25−50 mbar (190−380 mmHg) reduced the percentage of sweet cherries, strawberries, and artificially wounded table grapes that subsequently developed decay. Sweet cherries and strawberries were exposed to a pressure of 0.25, 0.50, and 0.75 atm for 4 h, or to 0.5 atm for 1, 2, or 4 h; table grape bunches to 0.25, 0.5, and 0.75 atm for 24 h and then they were inoculated by spraying with a conidial suspension (1×10^6 spores per mL) of *B. cinerea*. Following hypobaric treatment, the percentage decay was determined after sweet cherries were stored in the dark at 0°C for 14 days, followed by 7 days shelf life at 20°C; strawberries and table grapes were kept in the dark at 20°C for 3−10 days. A 4-h treatment of sweet cherries at 0.5 atm reduced the percentage of fruits that developed grey mold rot by 66−99.3%, and reduced total rots, including grey mold rot, brown rot, blue mold rot, and green rot, by 55−67%. During 4 days at 20°C following a 4-h treatment at 0.5 atm, the percentage of *Rhizopus* rot that developed on strawberries was reduced by 80%, grey mold rots by 22%, and total rots by 22%. Keeping grapes at 0.25 atm for 24 h decreased the percent infection that developed by 53% during 10 subsequent days at 20°C. A direct inhibition of mold growth due to the lowered $[CO_2]$ and $[O_2]$ present during the hypobaric treatment was excluded as a cause of the reduction in fungal decay because growth of *Bacillus cinerea* and *Micrococcus laxa* cultures was not inhibited during 3 and 8 days, respectively, at 23°C and 0.25 atm. A hyperbaric pressure of 1.5 atm had a similar antifungal action on sweet cherries and table grapes, indicating that the response is induced by a change in pressure and not exclusively by a decrease in pressure (Romanazzi et al., 2008). Likewise Hashmi et al. (2013) reported that a 4-h exposure to 0.5 atm reduced natural fungal decay on strawberries during four subsequent days at 20°C, as well as decay induced by inoculation of strawberries with *B. cinerea* or *R. stolonifer* spores. They found that the activities of defense-related enzymes such as phenylalanine ammonia lyase and chitinase peaked 12 h after the hypobaric treatment, and peroxidase increased immediately, whereas polyphenol oxidase, which does not suppress mold infections, was unaffected. The increase in defense-related enzyme activity, and the associated reduction in rot incidence caused by natural infection or spore inoculation, was not duplicated by applying 10% $[O_2]$ at 1 atm. This indicates that the inhibition of decay was caused by a change in pressure rather than a reduction in $[O_2]$. These studies suggest that the larger inhibition of microbial growth at a low pressure compared to that at 1 atm and the same O_2 tension may be caused by the induction of defense-related enzymes. Likewise, the opening of stomata at pressures lower than 0.75 atm may involve enzyme activation due to a pressure change.

In tests performed between 2011 and 2013 at the Institute of Navy Research in China, She and Zheng found that 40 different types of fresh vegetables and fruits kept at 6−20 kPa and 0−14°C for 16−48 h ("hypobaric short-period treatment") and then stored at the same cold temperature with the cartons overwrapped with plastic film that reduced water loss, exhibited a marked decrease in fungal decay and a much longer storage life compared to a control present in a plastic-wrapped box that was not treated at a hypobaric pressure. For example, 95% of cauliflower stored for 36 h at a low pressure was still edible after it subsequently was stored for 104 days in MAP, remaining white, with 35.8 mg/100 g of vitamin C, and only developing a few black mold spots. Control cauliflowers turned black within a few weeks in a cold room.

12.7 HYPOCHLOROUS ACID VAPOR

Sodium hypochlorite (NaOCl) was first employed as a wound disinfectant by Hueter in 1831, and as a hand disinfectant by Semmelweis in 1847. Koch confirmed its bactericidal activity in 1881 (WallhäuBer, 1988). A NaOCl solution was sprayed into air to control contamination in a Lancashire cotton mill in 1918, and in 1928, doctors used a fine mist of seawater containing NaOCl to sterilize air. Masterman (1938) demonstrated that dispersing a NaOCl aerosol into air-killed air-borne bacteria, and in 1941 reported that hypochlorous acid vapor (HOCl) generated when NaOCl solutions are acidified by atmospheric CO_2, is the active germicide produced in hypochlorite aerosols. Less than a 5- to 30-min exposure to a low HOCl vapor concentration can be 100% effective in killing air-borne bacteria, and as little as 0.5 ppm HOCl vapor in air was >99% effective in killing the air-borne influenza virus within 7.5−30 min (Edward and Lidwell, 1943).

Myeloperoxidase (MPOase), the most abundant protein in white blood cells (neutrophils), generates HOCl from H_2O_2 and Cl^- (Albrich et al., 1981; Foote et al., 1983; Kettle and Winterbourn, 1997; Suzuki et al., 2002), and the HOCl plays an important role in the host defense microbiocidal reactions (oxidative burst pathway) of PMNs after they engulf invading pathogens. The gross features of the reactions are similar in phagocytosing PMNs, the cell-free MPOase-H_2O_2-Cl^- system, and in the response to applied exogenous HOCl. When bacterial cells are exposed to HOCl, iron−sulfur−proteins, β-carotene, nucleotides, lipids, protein amino groups, and porphyrins are rapidly reacted, enzymes containing cysteine are inactivated, DNA replication is inhibited, protein unfolding and aggregation is promoted, and rapid irreversible oxidation of cytochrome and adenine nucleotides occurs (Albrich et al., 1981). The bactericidal action of HOCl results primarily from the pathogen's loss of energy-linked respiration due to destruction of cellular electron transport components and the adenine nucleotide pool (Knox et al., 1948). Hypochlorous acid's effectiveness as a disinfectant is in part due to its ability to rapidly penetrate through a microorganism's cell membrane (Bunce, 1990).

Hypochlorous acid (HOCl) is at least 100 times more effective than OCl^- as a sanitizer and is responsible for most of the germicidal effect of active chlorine (Cl^+) both in air and chlorinated water. HOCl is a weak acid ($K_A = 2.9 \times 10^{-8}M$):

$$HOCl \leftrightarrow H^+ + ClO^- \tag{12.1}$$

Free available chlorine refers to the hypochlorous acid form of chlorine in a solution. Total free chlorine is the sum of associated hypochlorous acid (HOCl) and hypochlorite ion (OCl^-) present in the solution. All free chlorine would be in the form of hypochlorous acid if the pH was low enough (Figure 12.4). Above pH 8.0, the amount of available chlorine present in the solution, and the concentration of

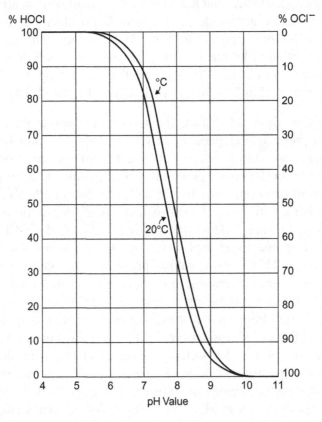

Figure 12.4 Distribution of hypochlorous acid and hypochloite ion in water at different pH values and temperatures. Morris, unpublished research at Harvard University, 1951.

HOCl vapor in air that has passed through the solution and equilibrated with it, decreases 10-fold for a 1-unit pH increase:

$$HOCl = (H^+)/(OCl^-)/K'_A \qquad (12.2)$$

where K'_A is the association constant for HOCl corrected for salting out. The dissolved chlorine gas (Cl_2) content in the solution (Figure 12.4) decreases 100-fold per unit increase in pH:

$$Cl_2 = (H^+)^2(Cl^-)(OCl^-)/K'_B K'_A \qquad (12.3)$$

where K'_B is the association constant for Cl_2 corrected for salting out.

Usually pH 6 to 7 chlorinated water is used as a liquid disinfectant because in that pH range essentially all chlorine in the solution is present as free-associated HOCl, and very little toxic Cl_2 gas or chlorine dioxide is present (Figure 12.4). US patent application 20080003171 entitled "Microbial control using hypochlorous acid vapor" claims that HOCl is "ineffective" above pH 8.5 (Smith et al., 2008). This is misleading because the HOCl is effective but very little is present at that pH.

Associated hypochlorous acid is volatile, unlike OCl^-. HOCl has not been used as a microbiocidal fumigant with fresh fruits, vegetables, or cut flowers because it is unstable and difficult to produce on-site at the precise low ppm vapor concentration needed to kill microorganisms without simultaneously damaging plant matter (Lister, 1952). Burg and Burg (1976) used the reaction between atmospheric CO_2 and a NaOCl solution to continuously generate hypochlorous acid vapor into air changes passing through laboratory NA, CA, and LP storage systems and determined the HOCl vapor concentration that killed various bacteria and fungi without damaging the host plant matter (also see Burg, 2004). HOCl vapor was continuously generated by adding between 0.1% and 2.5% alkaline sodium hypochlorite to water contained in a laboratory storage system's humidifier, in some instances buffering the mixture with Na^+ or K^+ salts of carbonate or bicarbonate, or titrating the solution down to a pH higher than 8.5 with HCl or nitric acid. Evaporated humidification water was continuously replaced, but the initial hypochlorite and additives were so concentrated that they did not need to be replenished for several months. Atmospheric CO_2 (0.036%), or the much lower CO_2 concentration present in incoming low-pressure air, was bubbled through the humidification solution, generating 0.35–10 ppm (μL/L vol:vol) hypo-chlorous acid vapor, measured colorimetrically by passing the chlori-nated air change through o-tolidene reagent. HOCl vapor prevented mold and bacterial growth at both atmospheric and subatmospheric pressure. The method by which the hypochlorite solution was brought into contact with air was not critical. Therefore, HOCl can be conve-niently generated in any system that uses a mechanical humidifier.

As rapidly as hypochlorous acid is lost from a NaOCl solution, it is regenerated from the large excess reservoir of hypochlorite ion present, causing the solution's alkalinity to decrease by 0.1–0.3 pH units dur-ing several weeks or months. After equilibration with CO_2 present in the incoming air change has occurred, the solution's pH stabilizes between 8.5 and 9.3 due to the buffering action of hypochlorous acid, which has a pK of 7.54.

To preserve electric neutrality in a hypochlorite solution that has equilibrated with incoming CO_2, the concentrations of the various ionic components must give rise to a total electric charge carried by cations that is equal and opposite to that carried by anions:

$$(Na^+) + (H^+) = K_1'q'(pCO_2)/(H^+) + 2\,K_1'K_2'q'(pCO_2)/(H^+)^2 \\ + C_AK_A'/[(H^+) + K_A'] + (Cl^-) + K_W'/(H^+) \quad (12.4)$$

where K_1' is the dissociation constant for carbonic acid, corrected for salting out according to the expressions $pK_1 = pK - 0.5\omega^{1/2}$ and $\omega = \tfrac{1}{2}$ $\sum m_1Z_1^2$; m_1 is the molality of the ion and Z_1 its valence; K_2' is the dis-sociation constant for bicarbonate corrected for salting out according to the expression $pK_2 = pK_2 - 1.1\omega^{1/2}$; q' is the Henry's Law constant for CO_2 corrected for salting out according to the expression log $(Q/Q') = 0.11\omega$, where $q = Q'/760$; pCO_2 is the partial pressure of CO_2 in air (mmHg); C_A is the hypochlorite concentration $(C_A = OCl^{-1} + HOCl)$; K_A' is the dissociation constant of hypochlor-ous acid corrected for salting out; and K_W' is the dissociation constant of water corrected for salting out. This equation predicts, and tests ver-ified, that sodium hypochlorite solutions ranging in concentration from 0.1% to 2.6% (weight/volume) equilibrate with atmospheric $[CO_2]$ at pH values in the range 8.5–9, and that when the $[CO_2]$ is increased or hypochlorite concentration decreased, the equilibrium pH is lower. When Na^+ or K^+ salts of bicarbonate or carbonate are added to a hypochlorite solution, a higher pH is maintained, the hypochlorite reservoir depletes more slowly, and a lower concentration of hypochlorous acid vapor is generated. If the pH decreases below 6.0, the solution becomes unstable and rapidly decomposes, generating large amounts of chlorine gas. During LP storage, the $[CO_2]$ partial pressure of air passing through a hypochlorite solution is much lower than it is at atmospheric pressure, in proportion to the pressure reduction. This elevates the solution's equilibrium pH and causes a lower concentration of hypochlorous acid vapor to be evolved for a longer duration of time. Cl^- is the only significant residue left on plant matter treated with HOCl vapor.

When conditions are controlled so that 0.5–9.1% of the available chlorine is present as associated volatilizable HOCl in the aqueous NaOCl solution, the remaining unassociated hypochlorite ions, which greatly predominate, are nonvolatile and held in reserve. The HOCl distributes itself between the solution's gas and liquid phases in accord with Henry's Law ($K_H^\circ = 730$ mol/kg bar; Holzwarth et al., 1984; Sander, 1999), and the air change continuously entrains a low concentration of the associated, nonionized HOCl. Simultaneously, additional volatilizable associated HOCl is created from the reservoir of unassociated chlorine-containing ions in a self-sustaining chemical reaction. The solution continues as a reservoir of associated HOCl over a long period of time with only a slight consumption of the hypochlorite ion. Evaporated humidification water must be continuously replaced, but the initial reservoir of hypochlorite and additives is so concentrated that it does not need to be replenished for weeks or months. As rapidly as hypochlorous acid vapor is lost from the solution it is regenerated from the large excess of hypochlorite ion present, causing the solution's alkalinity to gradually increase by 0.1–0.3 pH units. This slightly decreases the HOCl vapor concentration in the effluent air.

The ability to evolve HOCl vapor at a nearly constant rate for long periods of time is due to feedback controls that govern the availability of associated HOCl for entrainment in the carrier air. The pH of the aqueous solution is maintained sufficiently alkaline and above that corresponding to the pK_A of HOCl to continuously provide a trace amount of associated HOCl while inhibiting its loss by light-enhanced reactions and the generation of chlorine gas or chlorine dioxide. Below pH 8.5, hypochlorite solutions become unstable because the associated HOCl is converted to O_2, HCl, and $HClO_3$. Since HCl and $HClO_3$ are stronger acids than HOCl, these reactions acidify the solution even though a certain amount of acid is lost as HOCl vapor. When the pH decreases, the percentage of total available chlorine present as HOCl increases and therefore the rate of acidifying reactions also increases. This autocatalytic sequence, initiated below pH 8.5, results in the total decomposition of the solution to yield HCl, $HClO_3$, O_2, and HOCl vapor. Below pH 5, toxic Cl_2 gas forms and is released in progressively increasing amounts (Figure 12.4).

Aqueous solutions containing between 0.1% and 2.6% NaOCl can be adjusted so that after they equilibrate with atmospheric CO_2 or a much lower CO_2 concentration passing through an LP system's humidifier, the solution initially contains between 0.5% and 5% associated HOCl and yields between 0.5 and 9 ppm (μL/L vol:vol) of hypochlorous acid vapor. Fumigation with between 0.1 and 5.9 ppm HOCl vapor was 100% effective in preventing decay from developing on tomatoes, bananas, peppers, cucumbers, green beans, grapes, oranges, grapefruit, limes, avocados, mangos, lettuce, strawberries, and pineapples at storage pressures ranging from 30 to 760 mmHg for storage times varying from 14 to 150 days (Burg and Burg, 1976; Burg, 2004). Without HOCl treatment control plant matter developed 33–100% decay in a much shorter time. At 100 mmHg, a 0.5–1 ppm HOCl vapor concentration prevented growth of *C. gloeosporioides*, *Thielaviopsis paradoxa*, *Verticillium theobromae*, *B. cinerea*, *Sclerotinia sclerotiorum*, and *G. musarum* cultures on agar medium at 20°C. The rate of carrier airflow is not critical over a wide range, but above some limiting flow rate the NaOCl should fail to completely equilibrate with air bubbling through the solution. To completely equilibrate with atmospheric CO_2, a 5.25% sodium hypochlorite solution must acquire 425 cc of CO_2 per 100 cc of solution. This is the amount of CO_2 present in 42.5 cubic feet of atmospheric pressure air. Therefore, equilibrium with CO_2 is not completed for more than 1 day when 4 standard cubic feet per hour (SCFH) of air is bubbled through 500 mL of a freshly prepared 5.25% NaOCl solution having no excess alkali. An equilibrated 0.11% solution contains only 0.4 cc of CO_2 per 100 mL and requires much less time to equilibrate. If no CO_2 is present in the air change, the pH of the sodium hypochlorite solution increases and generation of HOCl vapor ceases. The most important factor stabilizing the solution and preventing its pH from rising due to the loss of HOCl vapor is continuous equilibration with CO_2 present in the air change (Burg and Burg, 1976).

Hypochlorous acid vapor is a strong oxidizing agent that corrodes many materials. Polyethylene, polypropylene, methacrylates, polytetrafluorethylene, fiberglass-reinforced plastics, 5000 and 6000 series aluminum, glass, 316 stainless steel, and cadmium-plated steel were not affected during a 2- to 6-month exposure to 5–15 ppm HOCl vapor (Burg and Burg, 1976). Brass and 303 stainless steel were slightly corroded, and carbon steel and copper strongly corroded. Titanium is not affected, but the vapor attacks acetal and polybutene pipes and fittings and corrodes nickel. Neoprene reacts with HOCl vapor without visible damage and should be replaced with Viton. Cellulose fibers slowly remove active Cl^+ from air, and plastic foam depletes HOCl vapor so readily from an air stream that it can serve as a highly effective HOCl filter. Activated carbon also effectively filters HOCl vapor. Corrosion can be avoided by careful selection of materials, modern protective coatings, and generating a very low HOCl vapor ppm concentration in the air phase.

Although hypochlorous acid is thermodynamically unstable, it is much more stable than other hypohalite acids, and unlike the others can be distilled and recovered without extensive decomposition. This allows HOCl vapor to persist in air for long enough to control microbial development. When the HOCl oxidative reaction destroys or corrodes materials it depletes available chlorine (Cl^+) from the air. HOCl vapor also decomposes when it contacts cardboard boxes or the commodity's surface, and at atmospheric pressure this causes the HOCl vapor concentration within boxes to be much lower than it is in the storage air. Lowering the pressure to 10 mmHg enhances the diffusion of HOCl vapor 76-fold (Figure 2.2), and this should decrease the HOCl gradient between the storage atmosphere and interior of a box. In a 20-ft Fruehauf hypobaric intermodal container operating at 3°C and 160 mmHg, 2.5 μL/L of hypochlorous acid vapor was measured in the storage atmosphere and 1.0–1.5 μL/L inside cardboard boxes containing Shasta strawberries.

HOCl vapor can be rapidly vacuum infiltrated into the intercellular system of stored plant matter present in a hypobaric chamber. Immediately after the commodity has been cooled, packed in boxes, placed in an LP warehouse or intermodal container, and the chamber has been evacuated, it can be vented with air passing through a NaOCl solution. Venting can be completed in 30 min without damaging plant matter (Section 2.6), and the chamber may be reevacuated immediately thereafter, or a short time later, to begin hypobaric storage in a Vivafresh warehouse or transportation in a VacuFresh container. After storage or transport has been completed, the chamber can again be vented with air containing HOCl vapor. During venting, the HOCl vapor concentration within and around the plant matter progressively approaches the atmospheric concentration of the HOCl vapor in the venting air. Simultaneously, the HOCl in the chamber is decreased by contact with the commodity and boxes. By adjusting the pH of the sodium hypochlorite solution, and varying the concentrations of sodium hypochlorite and carbonate/bicarbonate buffer, the solution can be made to generate a wide range of HOCl concentrations. A high enough concentration can be maintained for long enough to kill up to 100% of bacteria, molds, and viruses present on and within the stored plant matter. After venting with HOCl containing air, residual chlorinated vapor can be flushed from a VacuFresh intermodal container by opening a vacuum valve on the door and pressurizing the container's pneumatic-venturi air horn at 60 psig with an accessory oil-free air compressor. The compressed air will exhaust two air changes per minute (1520 cfm) and cleanse 99% of the HOCl from the container's atmosphere in 5 min. This avoids having to pass residual HOCl through the vacuum pump's suction port, which will remain closed until the vacuum pump is energized to reevacuate the chamber.

The EPA considers HOCl to be nontoxic to human and animal tissues and nondeleterious to the environment. The FDA has approved the use of high aqueous concentrations of HOCl in contact with fresh foods and cut flowers (Smith et al., 2008).

12.8 OZONE

In 2001, the US Food and Drug Administration (FDA) declared ozone a safe antimicrobial agent for the treatment, storage, and processing of foods with gaseous and aqueous phases which directly contact fruits and vegetables. Elford and Van den Ende (1941) reviewed early medical studies on the use of ozone as an aerial disinfectant. They demonstrated that 0.025 ppm killed *Streptococcus salivarius* within 90 min if the humidity was high enough, and in a moderately humid atmosphere 0.2 ppm killed *Bacillus prodigiosus*, *S. salivarius*, and *Staphylococcus albus*. Less than 1 ppm O_3 kills *E. coli* K12 (Hamlin and Champ, 1974), in 1 ppm >90% of *Streptococcus mitis* and *Staphylococcus epidermis* bacteria died during a 5-min exposure (Pelleu et al., 1974), and in 4 h 0.3–0.9 ppm ozone was 95% effective in killing *Proteus*, *P. aeruginosa*, *Serratia*, *Aspergillus fumigas*, and three strains of *Staphylococcus aureus* (Dyas et al., 1983). The most effective continuously applied aerial ozone concentration that consistently prevents bacterial and fungal decay without injuring plant matter usually is between 0.1 and 2.0 ppm (μL/L). This concentration range prevents growth and sporulation of *P. digitatum* and *P. italicum* on oranges and lemons (Palou et al., 2001, 2002, 2003); *M. fructicola*, *Mucor piriformis*, and *P. expansum* on peaches and table grapes (Palou et al., 2001, 2002); *B. cinerea* on Kiwi fruits (Minas et al., 2011), tomatoes (Tzortzakis et al., 2007a,b), strawberries (Nadas et al., 2003; Perez et al., 1999; Tzortzakis et al., 2007a), table grapes (Tzortzakis et al., 2007a), plums (Tzortzakis et al., 2007a), and blackberries (Barth et al., 1995); decay in persimmons (Salvador et al., 2006); and *Schlerotinia sclerotiorum* and *B. cinerea* on carrots (Sharpe et al., 2009). A concentration of 0.2–0.3 ppm ozone was 99.997% effective in killing *E. coli* 0157:H7 on grape tomatoes during 14 days at 10°C, *Salmonella* on table grapes during 14 days, and 99.999% effective with *Listeria monocytogenes* on table grapes during 21 days at 4°C   (www.producenews.com/welcome/9-news-section/story-cat/4624-4368). *E. coli* was reduced by up to 1.4 log units when 8–16 ppm of ozone was vacuum infiltrated into spinach during repressurization after vacuum cooling (Vurma et al., 2009). Higher ozone concentrations have been applied intermittently for short intervals during LP storage (Zhang et al., 2005a,b; Li et al., 2004; Li and Zhang, 2005; Wang et al., 2008; Han and Zhang, 2006) and at atmospheric pressure, with highly variable results due to commodity injury. Perez et al. (1999) found that 0.35 ppm ozone caused a 40% reduction in the emission of volatile esters responsible for strawberry odor. A prolonged exposure to >0.4 ppm can be fatal to humans, and government regulations stipulate that the working environment must contain no more than 0.1 ppm ozone.

Ozone has 3000 times the oxidizing potential of HOCl and readily reacts with organic materials, especially if they are unsaturated. It is corrosive to metals, except 300-series stainless steel, gold, platinum, 5000-series and 6061 aluminum, and titanium. Ozone reacts so readily with a horticultural commodity's organic surface and with fiberboard storage boxes that fumigation with 0.7 ppm [O_3] failed to control sporulation by molds because only 0.1 ppm reached the surface of oranges present in plastic bags packed in fiberboard cartons. Adequate penetration to maintain 0.6 ppm in the bags only occurred when oranges were loaded inside plastic boxes with large vents or open tops (Harding, 1968; Palou et al., 2003). The O_3-gradient between the storage atmosphere and interior of a box should be reduced in LP because a hypobaric pressure accelerates ozone's diffusive entry into the box (Figure 2.2; Eq. (3.1)).

Because of ozone's short half-life it must be continuously generated *in situ* by exposing O_2 to high-energy electrons (hve^{-1}) formed during a corona discharge (CD) or to photon quantum energy contained in <200-nm wavelength UV radiation. High-energy splits an O_2 molecule into oxygen atoms (O) that combine with molecular O_2 to form O_3:

$$O_2 + hv(<200 \text{ nM UV or e}^{-1}) \rightarrow 2(O)$$
$$(O) + O_2 \rightarrow O_3 \tag{12.5}$$

Lightning in the troposphere[2] and spark discharges in the high-voltage field between the electrodes of a CD ozone generator (Figure 12.5) create a plasma in which NO forms from atmospheric N_2 and O_2 and reacts with O_2 or O_3 to form NO_2, which serves as a primary source of O_2 atoms during ozone formation in the

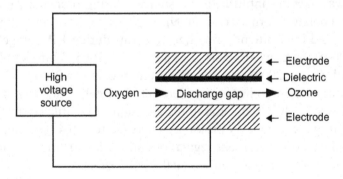

Figure 12.5 CD ozone generator.

[2]The troposphere extends to an average altitude of 17 km above the earth's surface, where the temperature is close to −68°C, the pressure near 110 mmHg, and the [O_2] concentration approximately 3%.

troposphere. Atomic (O) produced by the photolysis of ozone (Eq. (12.6)) reacts with water vapor to produce hydroxyl (OH) radicals, and NO_2 combines with water vapor to produce nitrous and nitric acids, which are highly corrosive when they form inside a CD generator's electrical discharge chamber. The hydroxyl radical can oxidize most of the chemicals found in the troposphere. To eliminate these undesirable reactions in a CD ozone generator, the air is filtered and dried to a dew point below $-60°C$ before it passes through the discharge gap. An effective method of cooling is required because 85–95% of the electrical energy supplied to a CD generator results in heat. CD generators are able to produce 3–6% ozone concentrations in air.

Ozone also can be generated by exposing air to 184.9 nm UV radiation produced by a low-pressure UVC or VUV mercury lamp. The UV spectrum is divided into VUV–vacuum ultraviolet at 10–200 nm; UVC at 100–280 nm; UVB at 280–315 nm; and UVA-blacklight at 315–400 nm radiation. All low-pressure Hg lamps emit radiation at both 184.9 and 253.7 nm (Figure 12.6), but UVB lamps do not produce ozone because they are constructed of a fused natural quartz that only transmits radiation >200 nm. UVB radiation is "germicidal" because 253.7 nm radiation is directly absorbed by DNA and is lethal to microorganisms. VUV and UVC lamps are constructed from an

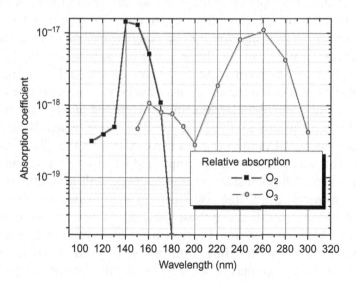

Figure 12.6 Absorption spectra of ozone and oxygen. Canadian Space Agency. In public domain.

artificial quartz synthesized from silica compounds, which transmits both 184.9 and 253.7 nm radiation. A photon of <200 nm light absorbed by an O_2 molecule splits it into two oxygen atoms (O) that can join with O_2 molecules to form O_3 (Eq. (12.6)). A photon of 200–300 nm light absorbed by an O_3 molecule splits it into an O_2 molecule plus an oxygen atom (O), which combines with an O_3 molecule to form two O_2 molecules:

$$O_3 + hv(253.7 \text{ nm}) = O_2 + (O)$$
$$O_3 + (O) = 2O_2$$

(12.6)

The equilibrium $[O_3]$ concentration produced by VUV and UVC low-pressure Hg lamps depends upon the rates of ozone creation and destruction by reactions described in Eqs. (12.4) and (12.5). UV absorption by O_2 and O_3 is directly proportional to the lamp's radiation intensity at 184.9 and 253.7 nm, and depends on the O_3 and O_2 concentrations and their photo-dissociation cross sections. Eighty percent of a UVC or VUV lamp's radiation is produced at 253.7 nm, only 20% at 184.9 nm (Figure 12.6). The largest $[O_3]$ concentration that can be emitted by a VUV or UVC lamp is a least 10-fold lower than the maximum concentration produced by a CD generator.

The Purfresh intermodal container ozone generating system, (Dick et al., 2008), "consists of a gas circulating means for producing a circulating stream of oxygen-containing gas within the container, a refrigeration unit for refrigerating the oxygen-containing gas, a CD ozone generator, an ozone injector for injecting ozone into the evaporator fan exhaust downstream of the evaporator, sensing means for determining the ozone concentration in the gas stream, and a controller for controlling the ozone injecting means to achieve a target ozone concentration in the circulating stream. In some embodiments, the container may also have means to dilute the injected ozone, which typically has a concentration >1 ppm in the gas mixture exiting the CD ozone generator. This dilution may be accomplished by mixing the ozonated gas with a volume of gas inside the container, or with a volume of air or CA gas introduced into the container. In many cases the appropriate rates of ozone generation are those that will achieve an ozone concentration of between 0.08 and 1.2 ppm by volume into an injected mixture having a volumetric flow rate between 1 and 20 cubic feet per hour. The invention may also include means for the controller to determine the status of the recirculating refrigeration evaporator fans, and reduce or shut-off ozone injection when a recirculating fan is off, as during a defrost cycle. Means also are provided to concentrate the ozone at a specific location in order to destroy ethylene in the circulating air before the ozone becomes diluted."

A VUV or UVC lamp can be used to produce ozone in a hypobaric warehouse since in a stationary situation the lamp will not be subjected to in-transit shock and vibration. Such lamps do not require dry atmospheric air to operate properly, they produce ozone at the same rate regardless of the dew point, do not form nitrous and nitric acids, and need less frequent cleaning and maintenance than a CD generator (Dimitriou, 1990). The life expectancy of VUV and UVC lamps is 10,000–20,000 h at a 25°C optimal temperature for ozone production. The lamp's irradiance decreases rapidly when it is first operated, but within a few hours stabilizes, and then decreases by an additional 50% during several years of usage. The ozone output of UVC or VUV lamps can be adjusted between 0% and 100% by an electronic dimmable ballast.

Due to in-transit shock and vibration, a CD generator rather than a UV light must be used to continuously produce ozone in a VacuFresh hypobaric intermodal container. Ozone production by a CD generator can be regulated by a controller, and the ozone-containing air introduced downstream of the pneumatic air horn (Figure 6.1) in the container's longitudinal duct (Figure 13.2-K). There is no need to locally concentrate the ozone to destroy recirculating ethylene, as the Purfresh system attempts to do, because hypobaric storage automatically lowers the ethylene concentration within plant matter, and low-pressure air changes prevent ethylene from accumulating in the storage air. None of the interior materials in VacuFresh containers are reacted by ozone, except the neoprene door seal, which can be replaced by an ozone-resistant Viton door seal.

Government regulations mandate that the ozone concentration in rarified air exhausted from a warehouse or an intermodal container must be lower than 0.1 ppm. A thin layer of activated carbon particles in a pleated configuration does not create a large enough pressure drop to significantly decrease a vacuum pump's efficiency (Gundel et al., 2002), and although a high RH interferes with this filter's efficiency, the RH is lowered to <50% before ozone reaches the filter since the low-pressure exhaust air warms to 25°C prior to entering the vacuum pump. Gas ballast air prevents residual ozone from increasing in the vacuum pump's oil when the pump compresses the low-pressure mixture back to atmospheric pressure.

Carrier Transicold withdrew its support for the Purfresh ozone-based system due to ozone's corrosivity on copper and rubber components in the Carrier Transicold refrigeration system (World Cargo News, April 2012, p. 18) (also see Section 11.9).

12.9 GERMICIDAL EFFECT OF IONIZING RADIATION

Using ionizing radiation to prevent microbial growth in foods has been under investigation since the late nineteenth century. Irradiation reaches microorganisms in surface, subsurface, and interior regions of fresh produce, and when it strikes bacteria and other microbes, its high-energy breaks chemical bonds in molecules that are vital for cell growth. The microbes die or can no longer multiply and cause spoilage.

Much of irradiation's germicidal effect results from the ionization of water to yield free radicals. Usually the maximum radiation dose that fresh commodities can tolerate without developing ripening abnormalities, loss of firmness, altered flavor, and increased suscepti-bility to mechanical injury is close to 2.25 kGy (Sommer and Fortlage, 1966). When spores are present a minimum dose of 1.75 kGy is required for effective inhibition of most types of posthar-vest fungi (Kader, 1986; Aziz and Moussa, 2002; Table 12.2). Between 0.15 and 0.5 kGy is effective with fresh cut lettuce (Hagenmaier and Baker, 1997). Growth of three species of citrus molds was delayed by 1.5 kGy, and 5 kGy killed these fungi (Ladaniya et al., 2003). In tests with pear pathogens, spores of *B. cinerea* and *P. expansum* were radiation sensitive and killed by 1 kGy, whereas 3 kGy was required to kill radiation-resistant *Alternaria tenuissima* and *Stemphylium botryosum* spores (Geweely and Nawar, 2006). Fungi usually are more resistant than vegetative bacterial cells to ionizing radiation (Chou et al., 1971). Table 12.2 lists the approxi-mate lethal radiation dose (kGy) required to kill insects, viruses, yeasts, mold, and bacterial vegetative cells and spores.

Table 12.2 Lethal Dose of Gamma Radiation

Organism	Approximate Lethal Dose (kGy)	Reference
Viruses	10–40	A
Yeasts	4–9	A
Molds		
Spores		
Alternaria tenuissima	3	B
Botrytis cinerea	1	B
Penicillium expansum	1	B
Stemphylium botryosum	3	B
Mycelial growth		
Alternaria alternata	1–2	C
Alternaria tenuissima	3	B
Aspergillus flavus	1–2	C
Botrytis cinerea	1	B
Curvularia geniculata	0.5–1	C
Penicillium expansum	2	B
Stemphylium botryosum	1.5	B
Trichoderma viride	0.5–1	C
Bacteria (cells of pathogens)		
Staphylococcus aureus	1.4–7.0	A
Salmonella spp.	3.7–4.8	A
Bacteria (cells of saprophytes)		
Gram negative		
Escherichia coli	1.0–2.3	A
Pseudomonas aeruginosa	1.6–2.3	A
Pseudomonas fluorescens	1.2–2.3	A
Enterobacter aerogenes	1.4–2.8	A
Gram positive		
Lactobacillus spp.	0.23–0.38	A
Streptococcus faecalis	1.7–8.8	A
Leuconostoc dextranicum	0.9	A
Sarcina lutea	3.7	
Bacterial spores		
Bacillus subtilis	12–18	A
Bacillus coagulans	10	A
Clostridium botulinum (A)	19–37	A
Clostridium botulinum (E)	15–18	A

(Continued)

Table 12.2 (Continued)

Organism	Approximate Lethal Dose (kGy)	Reference
Clostridium perfringens	3.1	A
Putrifactive anaerobe 3679	23–50	A
Bacillus stearothermophilus	10–17	A

Frazier and Westhoff, 1988 (A); Geweely and Nawar, 2006 (B); and Maity et al., 2011 (C).

Table 12.3 Effect of MSDD on Fungi and Bacteria (Lagunas-Solar et al., 2006)

Microorganism	Inoculums (spores, mL^{-1})	Media or Host Commodity	Result at 22–23°C
Botrytis cinerea	3.0×10^4	Petri dish	No colonies
		Table grapes	Infections delayed >20 days
		Berries	Infections delayed >10 days
Penicillium spp.	9.0×10^4	Petri dish	No colonies
Penicillium digitatum	4.0×10^6	Oranges, lemons	Infections delayed >45 days
Alternaria alternata	1.2×10^4	Petri dish	No colonies
Rhizopus spp.	1.8×10^4	Petri dish	No colonies
Escherichia coli 0157.H7	1.0×10^5	Petri dish	No colonies
Erwinia carotovora	n/a	Asparagus	No infections in 10–12 days
Salmonella typhimurium	1.0×10^5	Petri dish	No colonies
Staphylococcus aureus	1.0×10^5	Petri dish	No colonies

Permission granted via STM from Wiley & Sons to Elsevier for republication of Table 1.1. Summary results of MSDD disinfection studies—in Lagunas-Solar et al. (2006).

12.10 MSDD (SEE SECTIONS 11.6 AND 11.7 FOR A DESCRIPTION OF THE MSDD METHOD)

The biocidal effect of MSDD on fungi and bacteria was measured *in vitro* after inoculating 3-mm deep surface injuries on fruits and *in vivo* after spraying inoculum onto agar in petri dishes (Table 12.3). *E. carotovora* (soft rot), which infects asparagus by penetrating superficial wounds and bruises, did not develop on asparagus during 10–12 days after an MSDD treatment (no data was presented indicating the frequency and development of *E. carotovora* in control spears).

Cost-Effective LP Intermodal Container

The R&D cost is much larger for LP intermodal containers than for CA containers. Both systems are comprised of a module to control gas composition or pressure, an intermodal container, and a refrigerator, but the CA control module is installed in a "ready-made" conventional refrigerated intermodal container that does not need to be specially designed and tested. The LP control module and a refrigerator capable of controlling temperature in an evacuated space must be fitted into a specially designed and ISO-certified vacuum-tight intermodal tank container.

Hypobaric intermodal tank containers qualify under the "nonhazardous" International Maritime Dangerous Goods Code (IMDG) standard tank container classification since they do not carry cargo with a flashpoint of more than 61°C, there are no hazards from toxicity or corrosivity, and the equipment has a pressure rating of <20 psi, which is the lowest tank container pressure category. Lloyds of London approved the hypobaric designs, and the LP tanks were ISO tested and certified. There is no insurance or implosion problem, as suggested by Lougheed et al. (1977), the vacuum force is only 10% of the design load, and the major design stresses are caused by shock, vibration, the weight of the cargo and stacked containers, impact on the door if the cargo should break free, and mandated safety factors. These stresses arise to the same or a greater extent in all intermodal containers operating at an atmospheric or a positive pressure.

Grumman's Dormavac container was constructed from 216 aluminum I-beams welded together to form four $40''L \times 8''H \times 0.67''W$ double-walled panels, joined together with a door and bulkhead to form a square tank. Leakage was increased due to 2150 linear feet of porous weld seams in the wall panel's inner surface. The fabrication cost was elevated due to 4920 linear feet of structural welds, and the materials cost was excessive because the double-sided aluminum walls resulted in a 28,500 lbs container weight. A 40 ft cylindrical VacuFresh design eliminates leakage by decreasing the weld seams in the vacuum barrier to 115 linear feet, reduces the labor cost for fabrication since

Hypobaric Storage in Food Industry. DOI: http://dx.doi.org/10.1016/B978-0-12-419962-0.00013-9

there are only 1400 total feet of weld seams, and lowers the materials cost because the total weight of a 40 ft design, including equipment, is only 13,500 lbs. A 20 ft VacuFresh design reduces all of these factors by an additional 50%.

The 8 in. thickness of Grumman's Dormavac wall structure, and the 3 ft equipment bay depth, decreased the usable storage volume to 1500 ft^3. While this was sufficient space to load 40,000 lbs of dense cargo, the permissible payload was only 30,000 lbs because of statutory onboard ship and over-the-road gross weight restrictions. A 40 ft cylindrical VacuFresh container's usable volume also is 1500 ft^3, but the container's weight is less than half and its fabrication cost in South Africa is one-third as much as the cost and weight of Grumman's Dormavac design built in the United States. The entire storage space in a 20 ft VacuFresh hypobaric intermodal container can be loaded with dense cargo without exceeded the allowable over-the-road weight limitation.

Whether a VacuFresh hypobaric container is more costly than a CA container depends on how "cost" is defined. A grower or shipper wants to know his cost to rent the intermodal container or make a shipment, and for LP and CA that should be nearly the same and significantly less than for air transport. The "useful life" of a standard refrigerated intermodal container is seldom more than 7 years and can be less for CA because racking eventually destroys the leak integrity of the door seal. All intermodal containers are depreciated in 5–6 years, and CA containers often are financed during that term. A hypobaric intermodal container's strong and durable structure assures that it will realize the tank container industry's standard 20 year life expectancy, with a refurbishment required after 10 years at approximately 20% of the container's initial cost. Because of their long life expectancy, the increased fabrication cost of tank containers can be partially offset by financing them for a longer term at a lower yearly expense, while benefiting from a substantial tax advantage resulting from 5 to 6 years depreciation of the higher fabrication cost.

Lougheed et al. (1977, 1978) claimed that LP consumes excessive energy. This certainly was true of Dormavac containers, which needed a built-in 38 hp diesel engine and 20 kW generator to vaporize humidification water during a design-duration trip of 6 weeks. Without water recovery a 700 gallon on-board water tank would have been required, reducing the legally permissible over-the-road payload by 6000 lbs. To avoid this penalty, chilled water from a 50 gallon reservoir was continuously injected into a supercharged liquid-ring vacuum pump, cooling it and causing a low-temperature isothermal compression to condense humidification moisture, after which the condensed water was gravity drained back into the 50 gallon tank for reuse.

The power requirement was reduced in VacuFresh by replacing Grumman's mechanical humidification system with "metabolic" humidification (Chapter 7; Burg, 1987a,b), This eliminated the need for 3 kW of electric heat to vaporize water, 4.7 kW of refrigeration to reclaim the water, and allowed Dormavac's 7.5 hp supercharged liquid-ring vacuum pump to be replaced with a 3 hp oil-sealed gas-ballasted rotary-vein pump. The VacuFresh vacuum pump is mounted inside an insulated compartment, where a glycol-cooled radiator and thermostated glycol flow-controller remove its heat. By keeping the air temperature in the compartment at 25°C, water is prevented from condensing in the vacuum pump oil and other equipment exposed to the high-humidity process air. A brazed-plate heat exchanger and the jacketed refrigeration system remove heat transmitted through the VacuFresh tank's insulation before it enters the storage area, and the tank's leak-tight structure prevents ambient heat from infiltrating except in the controlled air changes. Due to the low pressure and density of the LP air changes, the pounds of air and kcal of sensible heat introduced per air change are much less in VacuFresh than in a conventional refrigerated intermodal container. No latent heat needs to be eliminated since the incoming air does not reach its dew point at the storage temperature after expanding and drying during entry (Sections 3.4 and 3.5). At an 80% RH and 38°C ambient condition, only 14 W (48 BTU/h) has to be removed flowing two air changes per hour through a 40 ft VacuFresh intermodal container operating at 10°C and 20 mmHg vs. 3444 W (12,120 BTU/h = 1 ton of refrigeration) to cool the same air change flowing at atmospheric pressure and 10°C through a conventional refrigerated intermodal container. In VacuFresh, the refrigeration compressor's capacity is balanced versus heat generated by the vacuum pump to insure that the compressor operates continuously in an unloaded state. This keeps the entire surface of the vacuum tank at a modulated constant temperature ± 0.2°C when the ambient temperature is 49°C. Additional heat is not needed in cold weather, and defrost is not required since secondary glycol coolant is used.

The cylindrical VacuFresh beam tank is self-supporting along its entire length. It is attached to the end-frames by aluminum mounting boxes welded to extensions of the tank barrel at the top, bottom, and both sides, pinned by huck bolts to T-1 tool steel mounting clips welded to the end-frames (Figure 1.2). Heat leakage through the mounting assembly is reduced by a fiberglass insulator sandwiched between each steel clip and aluminum box, and by using T-1 tool steel, which has half the thermal conductivity of carbon steel. To improve the tank's temperature constancy, the door-seal mates against the surface of a triangular aluminum extrusion through which refrigerated glycol is flowed (Figure 13.2-F).

In CA and NA systems, metabolic heat is removed by refrigeration, consuming power, but in LP most of the heat produced by respiration and fermentation is transferred by evaporative cooling and the water vapor is evacuated, independent of refrigeration (Chapter 7, Burg and Kosson, 1983). VacuFresh does not have energy-consuming, heat-producing evaporator fans. Instead ventilation is provided by a pneumatic air mover (Figure 8.1), which has no moving parts, produces no heat, consumes no additional power, and is operated by the pressure difference created by the vacuum pump between the ambient atmosphere and the interior of the vacuum tank. CA uses a 2.5 kW air compressor to produce N_2 gas; the original VacuFresh equipment package had a 2 kW vacuum pump.

VacuFresh has been redesigned to improve reliability, reduce cost and energy consumption, decrease commodity weight loss, and prevent water condensation on and water absorption by cardboard boxes. The vacuum breaker used in the original design has been replaced with a vacuum regulator (Section 5.3), and the new system (Figure 13.1) uses flow control devices described in US provisional patent application No. 6170501 (Burg, 2012) and Figure 13.1-#32, a scroll compressor (Figure 13.1-#24), energy saving improvements in the refrigeration system, an aluminum microchannel condenser coil (Figure 13.1-#28) and a shallow 3-phase 380/460 VAC 0.6 HP condenser fan (Figure 13.1-#30), a 2.0/2.4 HP vacuum pump with an integral oil reclaiming system (#16) in place of the original 3 HP pump and separate oil-reclaiming system, and a 1/2 HP glycol pump (Figure 13.1-#22) rather than the original 1.5 HP version. A spiral tank-stiffening ring (Figure 13.2) eliminates the straight pipes, short

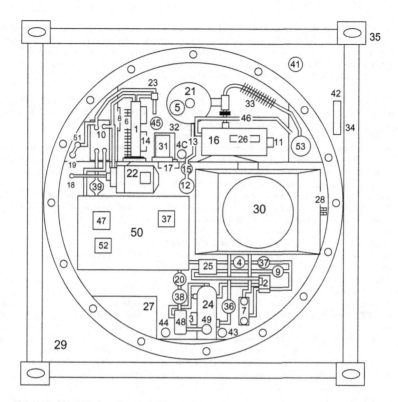

Figure 13.1 Revised VacuFresh equipment design.

1. *E-coated copper-fin/copper-tube radiator rated at 7200 BTU/h.*
2. *Heat exchanger (HX).*
3. *Electric cable connection to scroll compressor (#24).*
4. *Filter drier.*
5. *Pressure controller. Proportional 2-way normally closed (NC) valve with 0.025 in. diameter orifice and 24 VDC coil (Figure 6.1) operating responsive to a 7–28 VDC proportional valve driver. Mounted in parallel with a 2-way 24 VDC NC solenoid valve with a 0.06 in. orifice. Leakage through the proportional valve is reduced to 0.08 L/min by throttling air inflow through a 0.0039 in. orifice. Both valves operate with filtered air (#31). Flow entering above the diaphragm of the vacuum regulator (#21) is controlled by automatically adjusting current to the proportional valve between 0 and 85 mA in response to a signal from an absolute pressure transducer (#45).*
6. *Three (3) stainless steel 20 W/in.^{2} Finbar heaters. At 240 VAC, each Finbar provides 650 W per 17.625 in. length. Balanced power is taken between each 380/460 VAC 50/60 Hz phase and the neutral wire to obtain 1.8–2.2 kW of balanced heat from the three Finbars.*
7. *Receiver, with moisture sight-glass.*
8. *Air filter.*
9. *Injection expansion valve (EV).*
10. *Brazed-plate heat exchanger insulated with 0.5 in. thick ParaClad Spaceloft aerogel.*
11. *Vacuum pump oil filter.*
12. *Storage tank pressure sensing port (1/8 in. NPT).*
13. *Line from vacuum regulator (#21) to the tank pressure sensing port (#12).*
14. *Radiator fans, 24 VDC, 550 cfm at 1650 rpm. Powered by a 3-phase 320–574 VAC 47–63 Hz din-rail mounted supply delivering 24 VDC ± 2% at 240 A.*
15. *24 VDC NC brass vacuum solenoid valve.*
16. *Vacuum pump (230/400 V 50 Hz/460 V 60 Hz 3-phase motor, 2.0/2.4 HP at 50/60 Hz with standard gas ballast and oil filter, exhaust filter gauge, and oil level sensor. The pump starter cannot be energized until an*

(Continued)

NO thermal switch set to close above 12.8°C senses that the oil temperature is warm enough to start the pump without overloading the pump's motor. When the oil temperature initially is less than 12.8°C, the Finbar heaters (#6), fans (#14) mounted on the radiator (#1), and vacuum pump heaters (#26) turn on. Heaters turn off and the vacuum pump starts when the temperature switch closes. Pump is shock and vibration mounted. Pump emits a maximum of 7000 BTU/h of heat.

17. *Fork lift slot(s).*

18. *Connection to glycol surge tank. The vacuum pump cabinet is insulated with 0.5 in. thick ParaClad Spaceloft aerogel, leaving sufficient fore/aft space for the vacuum pump to move on its shock absorbers. Heat that penetrates into the glycol surge tank lowers the heat load on the radiator H/X (#1).*

19. *Glycol 1.5 in. supply to the VacuFresh tank's cooling spiral and door-frame duct.*

20. *Digital unloader solenoid valve. The compressor controller closes the unloader valve when the glycol temperature leaving heat exchanger #10, measured by temperature probe #51, is within 2.5°C of the set-point temperature. Then the controller "modulates" or cycles the unloader valve in an On/Off pattern during a 20 s period according to the capacity demand signal from the system controller, regulating the cold returning refrigerant flow from heat exchanger #10. The digital unloader solenoid valve may be replaced by using an R134a scroll compressor with a discharge dome temperature sensor and digital unloader valve (DUV) controlling the compressor capacity by disengaging the scroll wraps.*

21. *Vacuum regulator (1.5 in.) with external register (Figure 5.2G).*

22. *Glycol pump with 0.5 HP 3-phase 208–230/460 VAC TEFC motor.*

23. *Stainless steel reverse acting ambient air sensing thermostatic valve preset to control the upper radiator fan's air discharge between 18.3°C and 23.9°C, close to 21.1°C.*

24. *R-134a scroll compressor.*

25. *Economizer heat exchanger.*

26. *Three 225 W, 230 VAC, flexible silicone rubber heater strips bonded to the vacuum pump oil reclaiming case with RTV to preheat the pump's oil before starting it in cold weather. 240 VAC to the heaters is supplied using power taken between each 380/460 VAC 50/60 Hz phase and the neutral wire to obtain approximately 690 W of balanced heating. The oil temperature must exceed 12.8°C before the pump can be started without overloading its motor. The oil temperature is measured with a thermostat inserted in the pump's oil drain. Oil heating is assisted by the Finbar heaters (#6). The thermostat is set to close at >12.8°C, and circuitry is adjusted to allow the vacuum pump to turn on and the Finbar heaters (#6) and silicone rubber vacuum pump heaters turn off when the oil temperature exceeds 12.8°C.*

27. *Cable storage.*

28. *Aluminum microchannel condenser coil, providing 65,000 BTU/h at 95°F ambient and a 115°F condensing temperature. All-aluminum construction reduces galvanic action, and the coil is E-coated for a corrosive environment.*

29. *Steel end-frame reinforcing gusset allows use of standard iron end-frames (#34).*

30. *Axial fan (450 mm) with 3-phase 380/480 VAC 50/60 Hz motor providing 1370 rpm at 0.98 A. Maximum ΔP = 0.8 in. wc. Flowing 4500 cfm expect a 0.66 in. wc. Motor can have 2 speeds to automatically reduce power consumption when operating at a low heat load. Approximately 6.5 in. depth with guard grill. Maximum power input is 0.98 kW at 1.6 A. Draws air through grills from four sides to the back of the condenser coil (#28).*

31. *Spin-on hydraulic air filter with 3 μm canister. Aluminum head and cellulose filter media. Supplies 20–25°C filtered air to vacuum pump gas ballast (#16), thermal mass flow controller (#32), and "bleed" solenoid valve on the vacuum regulator (Figure 5.2-G).*

32. *Low-ΔP flow meter with control valve and readout/control module. The meter and valve are mounted behind the 3 μm spin-on air filter (#31), and the controller (#47) is in the electric cabinet (#40). Air to the flow meter and control valve from the spin-on air filter (#31) flows to the pneumatic air horn (Figure 6.1) through NC solenoid valve #40. Without power, the mass flow controller is normally closed. It provides sufficient flow to saturate the air in the VacuFresh container with all types of stored products after pressurizing the air horn and flowing through it. The system is shut off by NC solenoid valve #40.*

33. *Finned copper H/X to elevate the temperature of low-pressure air before it reaches vacuum regulator #21.*

34. *End-frame.*

35. *Corner casting.*

36. *Discharge pressure regulator.*

37. *390/460 VAC to 24 VAC transformer for control circuits.*

38. *Check valve.*

39. *Electronic expansion valve (EEV).*

40. *Normally closed 24 VDC 8210 brass 1/2 in. NPT ($C_v = 4$) solenoid valve discharging from thermal mass flow controller (#32), supplying air to the pneumatic air horn and an overpressure by-pass pop-off relief valve installed in the collar of the air horn (set to crack-open at 379 mbar).*

(Continued)

41. *Emergency glycol fill cap.*
42. *Sight glass indicating glycol level in the surge tank.*
43. *Glycol surge tank drain ball valve, 1/2 in.*
44. *1.5 in. FNPT NC ball valve to drain glycol.*
45. *Absolute pressure transducer.*
46. *Vacuum pump discharge air warms incoming low-pressure air to vacuum regulator #21 (Figure 5.2).*
47. *Panel mounted controller to regulate thermal mass flow controller (#32).*
48. *Oil separator.*
49. *Oil-return service valve.*
50. *Electric component box.*
51. *RTD temperature probe, inserted in stainless steel well, sensing glycol flowing into the VacuFresh tank cooling extrusion.*
52. *Digital Compressor Controller or microprocessor.*
53. *Vacuum pump suction line connection to floor vacuum suction line inside tank. When power is disconnected the vacuum pump's suction port closes and the pump vents to atmospheric pressure.*

radius elbows, entrance effects, and sudden enlargements associated with the circular stiffening rings and inter-ring flow connections in the original design (Figure 1.2). The spiral reduces the head needed to flow 30 gpm of glycol, from 49.5 to 28.8 ft, and this allows a 1/2 HP glycol pump to flow sufficient coolant to ensure $\pm 0.2°C$ end-to-end temperature uniformity on the tank's aluminum surface at an ambient temperature of 49°C.

The equipment layout in the redesigned VacuFresh equipment package is similar to that in any modern conventional intermodal container refrigeration system, except that two 3/4 HP evaporator fans and an E-coated aluminum-fin/copper-tube evaporator coil is replaced in VacuFresh by a secondary cooling system comprised of a brazed-plate heat exchanger (Figure 13.1-#10) and a spirally wrapped glycol-transporting rectangular aluminum extrusion welded to the tank's exterior surface (Figure 13.2-D). Eliminating the evaporator fans vacates sufficient space to accommodate the vacuum pump (Figure 13.1-#16) and other components required to create and control the low-pressure environment. There is insufficient space in the circular VacuFresh equipment package to install the copper-fin/copper-tube condenser and 3/4 HP 380/460 VAC condenser fan used in conventional intermodal container refrigeration systems. Instead, a 1.9 in. deep aluminum E-coated microchannel condenser (Figure 13.1-#28) and 3-phase 380/460 VAC 0.6-HP shallow evaporator fan (Figure 13.1-#30) are substituted to provide an equivalent condenser heat rejection capacity in a smaller space. The vacuum pump's 3-phase 2.4 HP, 50/60 Hz, 390/460 VAC motor continuously operates at full power pumping gas ballast and low-pressure air. The vacuum pump's

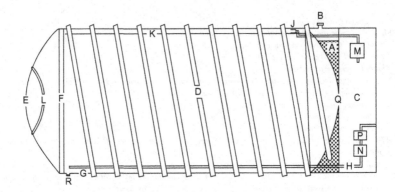

Figure 13.2 Diagram of a VacuFresh hypobaric intermodal tank container in which the pressure is controlled by a vacuum regulator and the cooling/stiffening extrusion is applied as a spiral wrap.
A—ethylene glycol refrigerant in surge tank;
B—vented surge tank cap;
C—temperature controlled equipment compartment (20–25° C);
D—spirally wrapped rectangular cooling and reinforcing rectangular extrusion;
E—front aluminum head (door);
F—neoprene or Viton door-seal seats against an aluminum triangular cooling and reinforcing extrusion;
G—5056-T6 marine aluminum circular vacuum tank suspended between end-frames by means of aluminum/fiberglass/T-1 steel mounting clips huck-bolted together and welded to the tank and end-frames (Figure 1.2);
H—floor duct in which holes are sized and spaced to provide uniform uptake along the entire duct length with less than a 5% pressure drop between the tank atmosphere and N;
J—pneumatic-venturi air mover (Figure 6.1) discharging into duct K;
K—longitudinal duct carrying venturi air mover discharge to front head E;
L—cooling/heating coil bonded to front door head;
M—thermal mass flow controller or proportional solenoid valve with remote thermal mass flow digital readout (Figure 13.1-#5);
N—absolute vacuum regulator (Figure 13.1-#21; Figure 5.2);
P—rotary gas-ballasted air-cooled oil-sealed vacuum pump;
Q—rear aluminum head cooled by glycol drawn from vented surge tank A;
R—vacuum-tight water drain.

energy consumption is partially compensated by eliminating nearly the entire sensible and condensation heat loads associated with air changes (Sections 3.4 and 3.5). Reducing the glycol pump's HP lowers the amount of "motor out, driven machine in" heat introduced into the flowing glycol, from 3280 BTU/h (1120 W) with a 1.5 HP motor to 1280 BTU/h (345 W) with a 1/2 HP motor, compared to 5040 BTU/h (1120 W) of "motor in, driven machine in" heat introduced into a conventional refrigerated intermodal container by two 3/4 HP evaporator fans (Avallone and Baumeister, 1987). The respiratory heat load in VacuFresh is decreased by an additional 90% due to the low [O$_2$] concentration at the storage pressure, and also because during hypobaric storage most of the respiratory heat is transferred by evaporative cooling and evacuated by the vacuum pump independent of refrigeration.

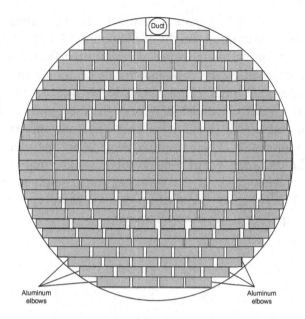

Aluminum
elbows

Aluminum
elbows

Figure 13.3 Box stacking arrangement for mango export in a 20 ft VacuFresh container. Irradiation approved box size (mm) = 370L × 275W × 90D. The box is recyclable, unable to absorb water, and its surface is coated with a radiation-reflecting substance. Screened vents in the box's sides prevent live insects from escaping and exchange heat to an adjacent air channel by evaporative cooling and to a lesser extent by convection. Exterior boxes, accounting for 26.6% of the total load, are primarily cooled by radiant transfer. To simplify loading, all boxes oriented horizontally instead of longitudinally are located along one side (left) or in the top row. Each mango is protected from in-transit bruising by an open-celled foam sleeve that is unable to absorb water. A Coppus 3 HP Jectair venturi air horn (Figure 10.2) discharges low-pressure air at sonic velocity from the rear to the front of the tank through a full-length longitudinal duct, and the air returns to the front through 2086 stagger-stacked boxes containing 7301 kg (16,106 lbs) of mangos (3.5 kg per box). The boxes are loaded and unloaded by means of an aluminum gravity roller conveyor inserted and withdrawn from the container at floor level. Exterior ends of outer boxes in the bottom three rows are supported by full-length 6 mm thick aluminum elbows welded to the tank shell in the vicinity of the external spiral stiffener. These angles allow boxes with heights of 90 mm (3.5 in.), 180 mm (7 in.), 270 mm (10.6 in.), or 360 mm (14.2 in.) to be loaded. The cargo weight is transferred into the spiral stiffener and 12 crescent-shaped floor supports. Transporting mangos for 30 days in a 20 ft VacuFresh container instead of for 2 days by air, from Nambia, India, to New York city, reduces CO2 emission by 97% per ton-mile, and the cost of fuel by 82%.

HP consumption is at least 37% lower in the redesigned unit compared to the original VacuFresh design.

The price of the redesigned VacuFresh tank container has been decreased by eliminating the material and labor costs associated with installing piping required for series glycol flow between circular stiffening rings, and deleting heavy aluminum gussets reinforcing each circular stiffening ring where its continuity was severed in the original design (Figure 1.2). The spirally wrapped stiffening/cooling ring is attached by a low-cost automated welding process (Figure 13.2-D).

Shelving in the original design has been deleted, eliminating its installation cost and a substantial weight of aluminum. A light-weight aluminum duct distributing low-pressure air exhausting from the pneumatic air horn is attached to the container roof's centerline in place of the original under-shelf duct (Figure 13.2-K; Figure 13.3). To further reduce weight, material, and labor costs, the meat rails and trolley in the original VacuFresh design have been removed and are offered as an optional accessory. These improvements offset the increase in cost due to inflation since VacuFresh prototypes were built in 1989 and reduce the current price of the container and equipment package to less than $100,000 after the start-up cost of prototype fixtures, jigs, and special equipment has been recovered.

CHAPTER 14

Storage Boxes

The waxed cardboard cartons used to prevent a loss of box strength during high humidity export shipments in intermodal sea containers have been banned in Europe because they cannot be recycled. They are replaced with laminated water-resistant recyclable paperboard boxes (Solidboard) that can be hydrocooled without absorbing water and losing strength, and by plastic boxes. Fiber box approved (FBA) cardboard boxes, such as International Paper's Climaseries or Interstate Container's Greencoat™, are impregnated with wax alternatives that make them recyclable, compostable, repulpable, and water resistant or waterproof. They have replaced waxed cardboard boxes in the United States. Any of these box types should prevent the weight loss caused by water condensation in nonwaxed cardboard boxes during hypobaric storage (Chapter 10). Vented open-topped waterproof plastic bins are now used to reduce floral water loss when roses and other flower types are stored in Vivafresh hypobaric warehouses.

A box's diffusive resistance to gas and vapor can be measured at atmospheric pressure by injecting a gas into the box, and determining the rate it escapes (Figure 14.1):

$$\tau = (V_{box}/A_{box})r_{box} \ln 2 \qquad (14.1)$$

where τ is the half-time (seconds) for the added gas to escape, V_{box} is the box's volume (cm^3), A_{box} the box's surface area (cm^2), and r_{box} the box's diffusive resistance to gas and vapor transport (cm/s). Measured in this manner, r_{box} is indicative of the box's resistance to water vapor since the binary diffusion coefficients of various gases and water vapor in air are closely similar. Equation (4.14) can be used to convert the resistance at atmospheric pressure to a resistance at any LP storage pressure. The kcal of heat generated by plant matter present in the box depends on the weight of plant matter in the box and the commodity's respiration rate at the storage pressure and temperature (Eq. (4.16)).

Hypobaric Storage in Food Industry. DOI: http://dx.doi.org/10.1016/B978-0-12-419962-0.00014-0

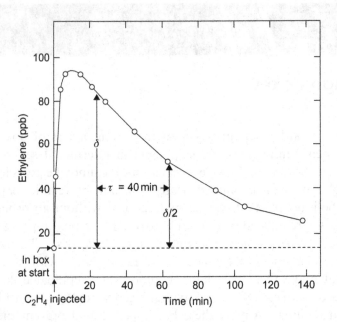

Figure 14.1 Outward diffusion of ethylene at 13°C and a barometric pressure of 560 mmHg from a 102 cm × 51 cm × 15 cm box containing 18.2 kg of carnation flowers, layers of newspaper, and a 1 mil polyethylene liner (150 cm × 185 cm) perforated with 80 1-mm diameter holes. Ethylene was injected one-third of the distance from each end of the box and subsequent samples, withdrawn from the center with a hypodermic syringe, were analyzed by gas chromatography. The ethylene concentration in the box decreased with a 40-min (2400-s) half-time (τ_b). The box resistance, computed from Eq. (14.1), is 597.4 s/cm at 560 mmHg (Burg, 2004).

Heat transfer by radiation and convection from the chamber wall to commodity present in storage boxes must be prevented by keeping the plant matter warmer than the vacuum chamber's wall in order to minimize commodity water loss during LP storage (Section 4.3), Then the only heat available to evaporate commodity water is the small amount of respiratory heat produced at a low pressure. This requirement is readily satisfied in a hypobaric warehouse or Dormavac intermodal container where the humidity is brought to supersaturation by a combination of mechanical humidification and commodity transpiration (see Chapter 6). Under these conditions a water-retentive plastic liner, such as a thin low-density polyethylene (LDPE) film inside storage boxes, easily keeps the commodity warmer than the chamber air and wall by restricting the escape of metabolic heat. The effectiveness of water-retentive wraps for this purpose is due to a horticultural commodity's low cuticular resistance to water vapor and high skin resistance to gas and vapor exchange (Ben-Yehoshua et al., 1985). For example, the surface resistance of Valencia oranges is similar for ethylene, CO_2, O_2, and very high, whereas its resistance to water is 60-fold lower. This large difference has been noted in numerous other types of fruits (Burg and Kosson, 1983; Cameron, 1982). O_2, CO_2, and ethylene's diffusivity cannot differ by more than ±12% in any medium because diffusivity depends on the square root of the molecular weight (Eq. (3.1)), but these gases have widely different air/water and air/oil partition coefficients (Section 3.1). Consequently, it is difficult to conceive of a mechanism other than an air phase that could transfer these gases though the fruit's surface with equal ease. Liquid water transport can occur in an aqueous phase in which gases diffusive 10^4 less readily than in air (Weis-Fogh, 1964). Evaporation at the outer cuticular surface creates a water potential difference that draws liquid water by mass flow from central to epidermal cells, through the cuticular membrane's liquid water phase, to the air/water interface (Schönherr, 1976; Schönherr and Schmidt, 1979). The transport resistances of a plastic liner and the fruit's cuticular surface summate because they act in series. If an appropriate plastic film is selected and, if necessary, the film is adequately perforated relative to the surface areas of the carton and commodity, and the commodity's weight and respiration rate, an amount of packaging resistance can be added that substantially elevates the low water vapor transport resistance without causing a significant increase in the high gas-transport resistance. The humidity inside the box can be elevated to saturate the air around the commodity without significantly changing the gaseous composition in the box's microclimate if the product of the wrap's resistance to water vapor and gas transport × the wrap's surface area is a multiple of the commodity's transpirational resistance × the total surface area of plant matter enclosed by the wrap, but significantly lower than the commodity's resistance to gas mass transport × the total commodity surface area within the wrap (Burg, 2004).

When LDPE is immersed in 27°C water for 48 h, it absorbs <0.01% of its weight as water. Therefore, a 1-mil LDPE liner should not cause a significant commodity weight loss by condensing water and releasing latent heat in the plastic (see Chapter 10).

A 10 mmHg storage pressure decreases convection by 89% (Burg and Kosson, 1982, 1983; Section 4.1); a low pressure inhibits heat conduction through micropores in the plant matter's cuticle and cardboard boxes (Section 4.4); and exterior boxes obstruct radiation between the

Table 14.1 Effect of Commodity Temperature on the Water Vapor and O₂ Partial Pressures Inside a Box at 0°C and a Total Pressure of 1.33 kPa (10 mmHg)

Commodity Temperature (°C)	Conditions Inside Box	
	Vapor Pressure (mmHg)	O₂ Partial Pressure (atm)
0	4.579	0.00149
0.5	4.750	0.00144
1	4.926	0.00140
2	5.294	0.00129
3	5.685	0.00119

chamber wall and interior boxes. Nevertheless, the temperature of roses in a 2°C Vivafresh warehouse does not increase significantly because a 10 mmHg pressure reduces the flower's respiration rate and heat production by 80–90% (Burg, 2004) and causes a 76-fold increase in diffusive water conductance through barrier air layers, stomata, the box, and protective wraps (Figure 2.2). A 0.36 mmHg water vapor partial-pressure gradient between the roses and microclimate in interior boxes, created by a 1°C increase in the flower temperature, is able to transfer all of a rose's respiratory heat by evaporative cooling. Flowing two air changes per hour, the elevated water vapor concentration in the box only lowers the O_2 partial pressure from 1.13 mmHg in the absence of flowers, to 1.06 mmHg with flowers present in the box (Burg, 2004; Table 14.1). An even smaller temperature increase in exterior boxes allows all respiratory heat to be dispelled by a combination of radiation, evaporative cooling, and convection. When tomatoes were stored in a Dormavac intermodal hypobaric container at 15°C and 10.67 kPa (80 mmHg), the fruit's temperature increased by 0.8°C in interior boxes and the 0.6 mmHg vapor pressure gradient between the commodity and box microclimate transferred water vapor through the box's surface at the same rate water evaporated from the tomatoes in response to their respiratory heat (Burg, 2004).

The situation is more complex in a VacuFresh hypobaric intermodal container because its venturi pneumatic air horn (Figure 6.1) emits ≥ 97% RH low-pressure air around boxes at the door end. This is not a high enough humidity to prevent an excessive commodity water loss at a low hypobaric pressure (Section 2.1). The water loss is minimal at the opposite end of the container because transpiration elevates the

RH to saturation as the recirculating air returns to the air horn through the stagger-stacked load. A water-retentive plastic box liner prevents evaporative cooling from lowering the commodity's temperature and causing radiation and convection to transfer latent heat from the chamber wall and storage atmosphere to plant matter in door-end boxes (Section 4.3). This stops the plant matter from acquiring environmental heat that would evaporate extra commodity water. At a 0°C storage temperature, a plastic wrap only needs to warm commodity present in a door-end box by 0.5–1.0°C to increase the RH in the box by 3.7–7.5% (Table 14.1). A temperature rise of this magnitude easily saturates the microclimate around plant matter in door-end boxes.

If the humidity initially is not saturated in a VacuFresh container, exterior boxes may evaporate extra water, cool, receive heat from the chamber wall, and continuously evaporate enough extra water to keep the humidity close to saturation. It may be necessary to inhibit radiation from the chamber wall to the exterior boxes to prevent this sequence from occurring before the pressure regulator and mass flow controller have sufficient time to saturate the chamber RH by adjusting the pumping speed and air-change rate (see Chapter 13). An unpainted aluminum chamber's interior has an emissivity of 0.2, and a box's natural emissivity is approximately 0.81 (Table 4.1). If both the interior of the chamber and the box's exterior surface are coated with white radiation-shielding pigment ($\varepsilon \approx 0.1$) the white surfaces will reduce radiant heat transfer from the chamber wall to the exterior box by 3.7-fold (Eq. (4.8)). Radiant transfer to plant matter ($\varepsilon = 0.96$) stored in the box will be reduced by an additional 8.4-fold if the box's inner surface is coated with a white radiation-shielding pigment, Radiation to the commodity from the interior of the box is reduced by an additional 2.1-fold if a Mylar® slip sheet ($\varepsilon = 0.2$) is inserted between the box's white inner surface and the commodity (Eq. (4.10)). The combination of a white chamber wall, white inner and outer box surfaces, and Mylar® reduces radiation to the commodity by 93%. However, inhibiting radiation from the chamber wall to the commodity also inhibits radiation in the reverse direction. Therefore, if the plant matter can be kept warmer than the chamber wall with an LDPE liner, that is a preferable solution because interior boxes will radiate a large portion of their respiratory heat to the chamber wall instead of transferring it by evaporative cooling.

Cellular expanded ethylene polyethylene (EPE) foam netting is often used to protect individual mango, tomato, and other fruits from shock and vibration injury during distribution. EPE is a waterproof closed-cell, cross-linked foam that takes up as little as 0.01% of its weight as water when it is immersed for 96 h at 27°C. EPE is unsuitable for use at a low pressure because its closed cells "explode" when the pressure is reduced. Noncross-linked PE foam netting should be satisfactory since it is open celled and virtually impermeable to water, taking up approximately 0.01% of its weight during a 48-h immersion in 27°C water.

CHAPTER *15*

Conclusions

The discovery of unanticipated processes operating during hypobaric storage has resulted in important patentable improvements to the LP method. The minimum $[O_2]$ concentration which does not cause low-$[O_2]$ injury in LP occurs at the RQ inversion point (IP) just as it does in CA, but pervaporation and enhanced gas and vapor diffusion displace the IP to a much lower $[O_2]$ concentration at the optimal hypobaric storage pressure. Water condensation in hypobaric warehouses can be eliminated and commodity weight loss reduced to a minimum by controlling the pressure with a vacuum regulator operating at 25°C, instead of a vacuum breaker. A regulator decreases commodity water loss in a hypobaric intermodal container by modulating the air-change rate to take advantage of the progressive reduction in metabolic heat generation that occurs during storage. Capillary condensation in conventional nonwaxed cardboard boxes is accelerated up to 100-fold at a low pressure, and releases latent heat that increases weight loss from stored plant matter. Waterproof plastic, waxed cardboard, or recyclable Solidboard, Climaseries or Greencoat storage boxes prevent water condensation from occurring, and if necessary a Mylar radiation reflecting liner can be installed between a box's inner surface and the stored commodity to reduce commodity water loss. The humidity in the redesigned VacuFresh container is computer controlled close to saturation in response to a thermal mass flow controller and wet- and dry-bulb $\pm 0.05°$ thermistors shielded from radiation with Mylar®.

A leak-free, flow-through hypobaric apparatus operating at a low pressure provides a longer storage life and less commodity water loss compared to any other storage system, it preserves plant matter that other methods do not benefit, and it prevents C_2H_4-induced responses, low-$[O_2]$ and high-$[CO_2]$ injury, directly inhibits bacterial and fungal growth, can vacuum-fumigate the commodity with HOCl vapor to kill molds and bacteria, and may kill surface insects. Power consumption

Hypobaric Storage in Food Industry. DOI: http://dx.doi.org/10.1016/B978-0-12-419962-0.00015-2

is lower in LP intermodal containers than in CA containers. LP consumes no chemicals, leaves no residue, is environmentally safe, and the cost to ship in an LP intermodal containers is much less than by air transport. LP is a logical choice to transport any type of plant matter that is presently exported with a loss of quality by air.

BIBLIOGRAPHY

Abe, Y., Kondoh, Y., 1989. Oxygen absorbers. In: Brody, A.L. (Ed.), Controlled/Modified Atmosphere/Vacuum Packaging of Foods. Food & Nutrition Press, Trumbull, CT, pp. 149–158.

Abeles, F.B., Morgan, P.W., Saltveit, M.E., 1992. Ethylene in Plant Biology, second ed. Academic Press, New York.

Abdul Raouf, U.M., Beuchal, L.R., Ammar, M.S., 1993. Survival and growth of *Escherechia coli* 0157:H7 on salad vegetables. Appl. Environ. Microbiol. 59, 1999–2006.

Acevedo, J.M., Centanin, L., Dekanty, A., Wappner, P., 2010. Oxygen sensing in *Drosophila*: multiple isoforms of the prolyl hydroxylase *fatiga* have different capacity to regulate HIFα/Sima. PLoS ONE (5/8):e12390.doi.1371/Journal.pone.0012310.

Ache, P., Bauer, H., Kollist, H., Al-Rasheid, K.A., Lautner, S., Hartung, W., et al., 2010. Stomatal action directly feeds back on leaf turgor: new insights into the regulation of the plant water status from non-invasive pressure probe measurements. Plant J. 62 (6), 1072–1082.

Affolter, M., Shilo, B.Z., 2000. Genetic control of branching morphogenesis during *Drosophila* tracheal development. Curr. Opin. Cell Biol. 12 (6), 731–735.

Aharoni, Y., Stewart, J.K., Hartsell, P.L., Young, D.K., 1979. Acetaldehyde—a potential fumigant for control of the green peach aphid on harvested head lettuce. J. Econ. Entomol. 72 (4), 493–495.

Aharoni, Y., Apelbaum, A., Copel, A., 1986. Use of reduced atmospheric pressure for control of the green peach aphid on harvested head lettuce. HortScience 21 (3), 469–470.

Akamine, E.K., Brewbaker, J.L., Budenhagen, I.W., Dollar, A.M., Hanaoha, M. 1771. Dosimetry, tolerance, shelf-life, extension related to disinfestation of tropical fruits by irradiation, Proceedings of a Panel, Honolulu, Hawaii, USA, 7–11 Dec. 1970. pp. 43–57, Report No. AGE-PL/422/7.

Akamine, E.K., Moy, J.H., 1983. Delay in postharvest ripening and senescence of fruits. In: Josephson, E.S., Peterson., M.S. (Eds.), Preservation of Food by Ionizing Radiation, vol. 3. CRC Press, Boca Raton, FL, p. 129.

Akita, S., Mitasaka, A., 1969. Studies on the difference of photosynthesis among species. II. Effect of oxygen-free air on photosynthesis. Proc. Crop Sci. Soc. Japan 38, 525–534.

Akita, S., Moss, D.N., 1973. The effect of an oxygen-free atmosphere on net photosynthesis and transpiration of barley (*Hordeum vulgare* L.) and wheat (*Triticum aestivum* L.) leaves. Plant Physiol. 52, 601–603.

Al-Qurashi, A.D., Matta, F.B., Garner, J.O., 2005. Effect of low-pressure storage (LPS) on Rabbiteye blueberry "premium" fruits. Met. Env. Arid Land Agric. Sci. 16 (2), 3–14.

Albrich, J.M., McArthy, C.A., Hurst, J.K., 1981. Biological reactivity of hypochlorous acid: implications for microbial mechanisms of leukocyte myeloperoxidase. Proc. Natl. Acad. Sci. USA. 78 (1), 210–214.

Alvarez, A.M., 1979. Effects of hypobaric storage on quality and disease incidence of papaya fruits. Phytopathology 69 (9), 1020.

Alvarez, A.M., 1980. Improved marketability of fresh papaya by shipment in hypobaric containers. HortScience 15 (4), 517–518.

An, D.S., Park, E., Lee, D.S., 2009. Effect of hypobaric packaging on respiration and quality of strawberry and curled lettuce. Postharvest Biol. Technol. 52 (1), 78–83.

Anon, 1974. A Feasibility Study of Low Pressure Storage. Horticultural Science Department and School of Engineering, University of Guelph, Guelph, ON, pp. 46.

Aoki, S., Watanabe, H., Sato, T., 1974. Extending the storage life of mushrooms cv. Matsutake by gamma irradiation. J. Food Sci. Tech. (Tokyo) 21 (6), 290–292.

Apelbaum, A., Barkai-Golan, R., 1977. Spore-germination and mycelial growth and sporulation of some pathogenic fungi under hypobaric pressure. Phytopathology 67, 400–403.

Apelbaum, A., Aharoni, Y., Tempkin-Gorodeiski, N., 1977a. Effects of subatmospheric pressure on the ripening processes of banana fruits. Trop. Agric. (Trinidad) 54, 39–46.

Apelbaum, A., Zauberman, G., Fuchs, Y., 1977b. Subatmospheric pressure storage of mango fruits. Sci. Hortic. 7, 153–160.

Apelbaum, A., Zauberman, G., Fuchs, Y., 1977c. Prolonging the storage life of avocado fruits by subatmospheric pressure. HortScience 12 (2), 113–117.

Aquino, S., 2011. Gamma radiation against toxigenic fungi in food, medicinal and aromatic herbs. Formatex 2011, 272–280.

Arèvalo-Galarza, L., Follett, P.A., 2011. Response of *Ceratitis capitata*, *Bactrocera dorsalis*, and *Bactrocera cucurbitae* (Diptera: Tephritidae) to metabolic stress disinfection and disinfestation treatment. J. Econ. Entomol. 104 (1), 75–80.

Arèvalo-Galarza, L., Newmann, G., Follet, P.A., 2010. Potential for Metabolic Stress Disinfection and Disinfestation (MSDD) treatment to disinfest commodities of white peach scale and other surface pests. Proc. Hawaiian Entomol. Soc. 42, 49–52.

Argenta, L.C., Mattheis, J.P., Fan, X., Finger, F.L., 2004. Production of volatile compounds by Fuji apples following exposure to high CO_2 or low O_2. Agric. Food Chem. 52, 5957–5963.

Asen, S., Norris, K.H., Stewart, R.N., 1971. Effect of pH and concentration of the anthocyanin-flavanol co-pigment complex on the color of "Better Times" roses. J. Am. Soc. Hort. Sci. 96, 770–773.

Avallone, E.A., Baumeister, T., 1987. Marks' Standard Handbook for Mechanical Engineers, 9th ed. Chapter 12: Table 12.4.31. 9th edition. Chapter 10. Table 12.4.31. McGraw-Hill, New York.

Awad, M., Oliveira, A.I., deCorrrea, D.L., 1975. The effect of ethephon, GA and partial vacuum on respiration in bananas (*Musa acuminata*). Rev. Agric., Piracicaba, Brazil 50, 109–113.

Aziz, N.H., Moussa, A.A., 2002. Influence of gamma-radiation on mycotoxin producing moulds and mycotoxis in fruits. Food Control 13, 281–288.

Back, E.A., Cotton, R.T., 1925. The use of vacuum for insect control. J. Agric. Res. 31, 1035–1041.

Bangerth, F., 1973. The effect of hypobaric storage on quality, physiology, and storage life of fruits, vegetables and cut flower. Gartenbauwissenschaft 38, 479–508.

Bangerth, F., 1974. Hypobaric storage of vegetables. Acta Hortic. 1 (38), 23–32.

Bangerth, F., 1977. The effect of different partial pressures of carbon dioxide, ethylene and oxygen on the ascorbic acid content of vegetables. Qual. Plant Foods Hum. Nutr. 27, 125–134.

Bangerth, F., 1984. Changes in the sensitivity for ethylene during storage of apple and banana fruits under hypobaric conditions. Sci. Hortic. (Amst.) 24, 151–163.

Bangerth, F., Song, J., Streif, J., 2012. Physiological impacts of fruit ripening and storage conditions on aroma volatile formation in apple and strawberry fruit. HortScience 7 (1), 4–10.

Banks, H.J., Annis, P.C., 1990. Comparative advantages of high CO_2 and low O_2 types of controlled atmospheres for grain storage. In: Calderon, M., Barkai-Golan, R. (Eds.), Food Preservation by Modified Atmospheres. CRC Press, Boca Raton, FL, pp. 93–122.

Bare, C.D., 1948. The effect of prolonged exposure to high vacuum on stored product insects. J. Econ. Entomol. 41, 109–110.

Barker, J., el-Saifi, A.F., 1953. Studies on the respiratory and carbohydrate metabolism of plant tissues. III. Experimental studies of the formation of CO_2 and of change in lactic acid and other products in potato tubers in air following anaerobic conditions. Proc. R. Soc. London, Ser. B 140, 508–522.

Barth, M.M., Zhou, C., Mercier, J., Payne, A., 1995. Ozone storage effects on anthocyanin content and fungal growth in blackberries. J. Food Sci. 60 (6), 1286–1288.

Bellido, J., 2006. Ozone measurement systems: associated instrumentation and calibration. Tethys 3, 55–58.

Ben-Yehoshua, S., 1986. Transpiration, water stress, and gas exchange of fruits and vegetables. In: Weichmann, J. (Ed.), Postharvest Physiology of Vegetables. Marcel Dekker, New York, NY, Chapter II F.

Ben-Yehoshua, S., Burg, S.P., Young, R., 1985. Resistance of citrus fruit to C_2H_4, O_2 and CO_2 mass transport. J. Am. Soc. Hort. Sci. 105, 868–872.

Bender, R.J., Brecht, J.K., Sargent, S.A., Huber, D.J., 2000. Mango tolerance to reduced oxygen levels in controlled atmosphere. J. Am. Soc. Hort. Sci. 125 (6), 707–713.

Bhambhani, H.J., 1956. Responses of pests to fumigation. VI.—Water losses and the mortality of *Calandra* spp. at reduced pressures. Bull. Entomol. Res. 47, 749–753.

Blackman, F.F., 1953. Respiratory drifts. In: James, W.O. (Ed.), Plant Respiration. Oxford at the Clarendon Press, London, pp. 40–62.

Borecka, H., Olak, J., 1978. The effect of hypobaric storage conditions on the growth and sporulation of some pathogenic fungi. Fruit Sci. Rep. (Poland) 5, 39–41.

Bowyer, J.W., Leegood, R.C., 1997. Photosynthesis. In: Day, P.M., Harbone, J. (Eds.), Plant Biochemistry. Academic Press (an imprint of Elsevier).

Bramlage, W.J., Couey, H.M., 1965. Gamma radiation of fruits to extend market life. Marketing Research Report 717. USDA, Washington, DC.

Bramlage, W.J., Lipton, W.J., 1965. Gamma radiation of vegetables to extend market life. Marketing Research Report 703. USDA, Washington, DC.

Bredmose, N.B., Nielsen, K.L., 2009. Controlled atmosphere storage at high CO_2 and low O_2 levels affects stomatal conductance and influences root formation in *Kalanchoe* cuttings. Sci. Hortic. 122 (1), 91–95.

Brisker, H.E., 1980. Delaying Grapefruit Harvest: Hormonal and Enzymatic Control by Endogenous and Applied Regulators (Ph.D. dissertation). The Hebrew University of Jerusalem, Israel.

Brouwers, H.J.H., 1990. An improved tangency condition for fog formation in cooler-condensers. Int. J. Heat Mass Transfer 34 (8), 2387–2394.

Brouwers, H.J.H., 1992a. Film models for transport phenomena with fog formation: the fog film model. Int. J. Heat Mass Transfer 35 (1), 13–38.

Brouwers, H.J.H., 1992b. A film model for heat and mass transfer with fog formation. Chem. Eng. Sci. 47 (12), 3023–3036.

Brouwers, H.J.H., Chesters, A.K., 1992. Film models for transport phenomena with fog formation: the classical film model. Int. J. Heat Mass Transfer 35 (1), 1–11.

Buchsbaum, R., 1948. Animals Without Backbones. University of Chicago Press, Chicago.

Buck, J.B., 1962. Some physical aspects of insect respiration. Annu. Rev. Entomol. 7, 27–56.

Bunce, N.J., 1990. Environmental Chemistry. Wuerz Publishing, Ltd., Winnipeg, Canada.

Burg, S.P., 1967. Method for storing fruit. US Patent 3,333,967 and US Patent Reissue Re. 28,995 (1976).

Burg, S.P., 1973. Hypobaric storage of flowers. HortScience 8, 202–205.

Burg, S.P., 1975. Hypobaric storage and transportation of fresh fruits and vegetables. In: Haard, N.F., Salunkhe, D.K. (Eds.), Symposium: Postharvest Biology and Handling of Fruits and Vegetables. AVI Publishing Co., Westport, CT, pp. 172–188.

Burg, S.P., 1976. Low temperature hypobaric storage of metabolically active matter. US Patents 3,958,028 and 4,061,483.

Burg, S.P., 1987a. Hypobaric storage of respiring plant matter without supplementary humidification. US Patent 4,685,305.

Burg, S.P., 1987b. Hypobaric storage of non-respiring animal matter without supplementary humidification. US Patent 4,655,048.

Burg, S.P., 1990. Theory and practice of hypobaric storage. In: Calderon, M., Barkai-Golan, R. (Eds.), Food Preservation by Modified Atmospheres. CRC Press, Boca Raton, FL, pp. 353–372.

Burg, S.P., 2004. Postharvest Physiology and Hypobaric Storage of Fresh Produce. CAB International, Wallingford, Oxfordshire, UK, pp. 654.

Burg, S.P., 2005. Experimental errors in hypobaric laboratory research. Controlled Atmosphere Research Conference. Michigan State University, East Lansing, Michigan, USA, July 5–10.

Burg, S.P., 2012. Controlled and correlated method and apparatus to limit water loss from fresh plant matter during hypobaric storage and transport. Provisional Patent Application No. 61/705016, serial No. 14/033651, filed 9/24/2012. Non-provisional patent application Serial number 61/705016 filed 9/24/2013, approved by the USPTO on 2/23/14. Will issue as soon as the filing fee is paid. Assigned on 10/1/13 to American Management Group, LLC, Fort Washington, PA.

Burg, S.P., Burg, E.A., 1965. Gas exchange in fruits. Physiol. Plant. 18, 870–884.

Burg, S.P., Burg, E.A., 1966. Fruit storage at subatmospheric pressures. Science 152, 314–315.

Burg, S.P., Burg, E.A., 1976. Prevention of microbial growth. Canadian Patent 997,532.

Burg, S.P., Kosson, R.L., 1983. Metabolism, heat transfer and water loss under hypobaric conditions. In: Lieberman, M. (Ed.), Postharvest Physiology and Crop Preservation. Plenum Press, New York, NY, pp. 399–424.

Burg, S.P., Zheng, X., 2007. Summary of hypobaric research in China and the West. J. Refrig. 28 (2), 1–7 (in Chinese).

Burg, S.P., Zheng, X., 2009. Experimental errors in laboratory hypobaric research and answer. Chinese Assoc. Refrig. [A].

Burg, S.P., Bothel, R.D., Bothel, J.C., 2009. System and methods for controlled pervaporation in horticultural cellular tissue. Provisional Patent Application Serial No. 61/170,506, filed 4/17/2009. Non-provisional patent publication number 2010/0267144A1, serial number 12/760,776, filed 4/15/2010. Burg assigned his share in the patent to Atlas Bimetals Labs, Inc., Washington State, USA.

Burkett, B.N., Schneiderman, H.A., 1974. Roles of oxygen and carbon dioxide in the control of spiracular function in *cercropia* pupae. Biol. Bull. 147 (2), 274–293.

Bursell, E., 1957. Spiracular control of water loss in the tsetse fly. Proc. R. Entomol. Soc. London (A) 32, 21–29.

Burton, W.G., 1982. Postharvest Physiology of Food Crops. Longman, London, pp. 339.

Calderon, M., Navarro, S., 1968. Sensitivity of three stored-product species exposed to different low pressures. Nature 218, 190.

Calderon, M., Navarro, S., 1979. Increased toxicity of low-oxygen atmospheres supplemented with carbon dioxide on *Tribolium castaneum* adults. Entomol. Exp. Appl. 25, 39.

Calderon, M., Navarro, S., Donahaye, E., 1966. The effect of low pressures on the mortality of six stored product insect species. J. Stored Prod. Res. 2, 135–140.

Cameron, A.C., 1982. Gas Diffusion in Bulky Organs (Ph.D. thesis), University of California, Davis, CA.

Cao, Z.-m., Zang, P., Wang, L., 2004. Study on hypobaric storage and biochemical changes in postharvest Dongzao jujube. Storage Process 4 (4), 6–7.

Cao, Z.-m., Zhang, P., Wang, L., Gao, F., 2005. Effect of hypobaric storage on physiological and biochemical changes of Dong Jujube. Food Sci. 26 (10), 250–252.

Carpenter, A., Potter, M., 1994. Controlled atmospheres. In: Sharp, J.L., Hallman, G.J. (Eds.), Quarantine Treatments for Pests of Foods. Westview Press, Boulder/San Francisco/Oxford, pp. 171–198.

Centanin, L., Dekanty, A., Romero, N., Irisarri, M., Gorr, T.A., Wappner, P., 2008. Cell autonomy of HIF effects in *Drosophila*: tracheal cells sense hypoxia and induce terminal branch sprouting. Dev. Cell 14 (4), 547–558.

Centanin, L., Gorr, T.A., Wappner, P., 2010. Tracheal remodeling in response to hypoxia. J. Insect Physiol. 56 (5), 447–454.

Chang, J., Zhang, P., Wang, L., Tian, L., 2004. Effects of hypobaric storage on storage physiology of peach. Food Sci. 2004-01.

Chang, Y.-p., 2002. The developing and research foreground on new technology of hypobaric storage. Mach. Cereals Oil Food Process. 2, 8–9 (in Chinese).

Chang, Y.-p., Wang, R.-f., 2003. Physiological and biochemical study on hypobaric storage of Chinese jujube Dongzao. Food Sci. 24 (12), 135–137 (in Chinese).

Chao, Z.-m., Zhang, P., Wang, L., 2004. Study on hypobaric storage and biochemical changes in postharvest Dongzao Jujube. Storage Process 2004-04.

Chapman, R.F., 1998. The Insects. Structure and Function, fourth ed. Cambridge University Press, Cambridge, UK, pp. 770.

Chau, K.F., 1981. A Study on Anthracnose Caused by *Colleotrichum gloeosporiodes* Penz. on Papaya Fruit (*Carica papaya* L.) Under Normal and Low Pressure Storage (Thesis), University of Hawaii, pp. 130.

Chau, K.F., Alvarez, A.M., 1983. Effects of low-pressure storage on *Colletotrichum gloeosporioides* and postharvest infection of papaya. HortScience 18, 953.

Chen, H., Yang, H., Gao, H., Long, J., Yao, F., Fang, X., et al., 2013a. Effect of hypobaric storage on quality, antioxidant enzyme and antioxidant capability of the Chinese bayberry fruits. Chem. Central J. 7, 4.

Chen, H., Ling, J., Wu, F., Zhang, L., Sun, Z., Yang, H., 2013b. Effect of hypobaric storage on flesh lignification, active oxygen metabolism, and relayed enzyme activities in bamboo shoots. LWT—Food Sci. Technol. 51 (1), 190–195.

Chen, W.-x., Gao, H.-y., Chen, H.-j., Zhou, Y.-j., Yang, J.-t., 2004a. Changes of physiological and biochemical indexes under condition of hypobaric storage of Juicy Peach. Storage Process 46, 16–18 (in Chinese).

Chen, W.-x., Gao, H.-y., Mao, J.-l., Chen, H.-j., Zhou, Y.-j., 2004b. Research on technology of storage for hypobaric storage of Huanghua pear. Food Sci. 2004-11.

Chen, W.-x., Zhou, Y.-j., Chen, H.-j., Mao, J.-l., Gao, H.-y., 2005. Research on the technology of storage for hypobaric storage of bamboo shots. Food Sci. Tech. 2005-10.

Chen, W.-x., Gao, H.-y., Zhou, Y.-J., Chen, H.-j., Mao, J.-l., Song, L.-l., et al., 2007. Study of programmed hypobaric storage of "Cuiguan" pears. ISHS Acta Horticultuae 804. Europe-Asia Symposium on Quality Management in Postharvest Systems. Eurasia 2007.

Chen, Z., Marshall, S., White, M.S., Robinson, W.H., 2005. Low-pressure vacuum to control larvae of Hylotrupes bajulus (Coleoptera: Cerambycidae). In: Lee, C.-Y., Robinson, W.H. (Eds.), Proceedings of the Fifth International Conference on Urban Pests. Perniagann Ph'ng @ P&Y Design Network, Malaysia, pp. 325–329.

Chen, Z., White, M.S., Robinson, W.H., 2006. Preliminary evaluation of vacuum to control wood-boring insects in raw wood packaging materials. Forest Prod. J. 56 (7,8), July-August.

Chen, Z., White, S., Keena, M.A., Poland, T.M., Clark, E.L., 2008. Evaluation of vacuum technology to kill larvae of the Asian longhorned beetle, Anoplophora glabripennis (Coleoptera: Cerambycidae) and the emerald ash borer, Agrilus planipennis (Cleoptera: Buprestidae), in wood. Forest Prod. J. 58 (11), 87–95.

Chou, T.W., Salunkhe, D.K., Singh, B., 1971. Effect of gamma irradiation on Penicillium expansum h. III-on nucleic acid metabolism. Rad. Biol. 1, 329–334.

Cochrane, V.W., 1958. Physiology of Fungi. John Wiley & Sons, New York, NY.

Collander, R., 1937. The permeability of plant protoplasts to non-electrolytes. Trans. Faraday Soc. 33, 985–990.

Cominelli, E., Galbiati, M., Vavasseur, A., Conti, L., Sala, T., Vuyisteke, M., et al., 2005. A guard-cell-specific MYB transcription factor regulates stomatal movements and drought tolerance. Curr. Biol. 15 (13), 1196–1200.

Coorts, G.D., Gartner, J.B., McCollum, J.P., 1965. Effect of senescence and preservative on respiration in cut flowers of Rosa hybrida, "Velvet Times". Proc. J. Am. Soc. Hort. Sci. 86, 779–780.

Corey, K.A., 2000. Plant responses to rarified atmospheres. In: Mars Greenhouses: Concepts and Challenges. Proceedings from a 1999 Workshop. NASA Technical Memorandum 2000-20857, pp. 48–57.

Corey, K.A., Barta, D.S., Wheeler, R.M., 2002. Toward Martian agriculture—responses of plants to hypobaria. Life Support Biosph. Sci. 8, 103–114.

Couchat, P., 1977. Effet de l'oxygèn sur la transpiration. CR Acad. Sci. Paris Ser. D 285, 1303–1306.

Couchat, P., Lascève, G., 1980. Intervention de l'oxygèn atmosphérique sur la response hydrique du tournesol à une variationbrutale de temperature. CR Acad. Sci. Paris Ser. D 290, 271–274.

Couchat, P., Lasceve, G., Audouin, P., 1982. Dark stomatal movement in sunflowers in response to illumination under nitrogen. Plant Physiol. 69, 762–765.

Couey, H.M., Wells, J.M., 1970. Low-oxygen or high carbon dioxide atmospheres to control postharvest decay of strawberries. Phytopathology 60, 47–49.

Couey, H.M., Follstad, M.N., Uota, M., 1966. Low-oxygen atmospheres for control of postharvest decay of fresh strawberries. Phytopathology 56 (12), 1339–1341.

Covey, R.P., 1970. Effect of oxygen tension on the growth of Phytophthora cactorum. Phytopathology 60, 358–359.

Crozier, W.J., Pincus, G., Zahl, P.A., 1935. The resistance of Drosophila to alcohol. J. Gen. Physiol. 19 (3), 523–557.

Csonka, L.N., 1989. Physiological and genetic responses of bacteria to osmotic stress. Microbial. Rev. March, 121–147.

Davenport, T.L., 2007. Low pressure conditions kill quarantine insects while maintaining optimum tropical quality during shipments. CRIS Project Report. <www.reeis.USDA.gov/web/crisprojectpages/211739.html>.

Davenport, T.L., Burg, S.P., White, T.L., 2006. Optimal low-pressure conditions for long-term storage of fresh commodities kill Caribbean fruit fly eggs and larvae. HortTechnology 16 (1), 98–104.

Davenport, T.L., Burg, S.P., Follet, P., 2008. A lab scale low-pressure chamber system for conducting hypobaric research. HortScience 43, 1168 (Abstract).

Davis, P.L., Roe, B., Bruemmer, J.H., 1973. Biochemical changes in citrus fruits during controlled atmosphere storage. J. Food Sci. 38, 225–229.

Denny, M.W., 1993. Air and Water. The Biology and Physics of Life's Media. Princeton University Press, Princeton, NJ.

Dharkar, S.D., Savagaon, K.A., Spirangarajan, A.N., Sreenivasan, A., 1966. Irradiation of mangos. I. Radiation-induced delay in ripening of Alphonso mangos. J. Food Sci. 31, 863–869.

Dick, P., Cope, D., Wang, H., Hoobler, R., Weber, M., Volondin, A., 2008. Shipping Container Ozonation System. International Publication Number WO 2008/082452 A-1.

Dilley, D.R., 1977. The hypobaric concept for controlled atmosphere storage. In: Dewey, D.H. (Ed.), Controlled Atmospheres for the Storage and Transport of Perishable Agricultural Commodities. Proceedings of the Second National CA Research Conference, April 5–7, 1977. Michigan State University, pp. 29–37.

Dilley, D.R., 1978. Approaches to maintenance of postharvest integrity. J. Food Biochem. 2, 235–242.

Dilley, D.R., Dewey, D.H., 1973. Hypobaric storage of pomological fruits. NEM-27 Annual Report.

Dilley, D.R., Carpenter, W.J., Burg, S.P., 1975. Principles and application of hypobaric storage of cut flowers. Acta Hortic. 41, 249–268.

Dilley, D.R., Kuai, J., Poneleit, L., Zhu, Y., Pekker, Y., Wilson, I.D., et al., 1993. Purification and characterization of ACC oxidase and its expression during ripening in apple fruit. In: Pech, J.C., Latché, A., Balagué, C. (Eds.), Cellular and Molecular Aspects of the Plant Hormone Ethylene. Kluwer Academic Publishers, Dordrecht, pp. 46–52.

Dimitriou, M.A., 1990. Design Guidance Manual for Ozone Systems. International Ozone Association, Pan American Committee, Norwalk, CT.

Dumas, T., Buckland, C.T., Monro, H.A.U., 1969. The respiration of insects at reduced pressures. II. The uptake of oxygen by Tenebroides mauritanicus. Entomol. Exp. Appl. 12, 389–402.

Durkin, D., 1992. Pre-storage hydrating solution effects on flower life of roses after dry storage. HortScience 27 (6616).

Dyas, A., Boughton, B.J., Das, B.C., 1983. Ozone killing action against bacterial and fungal species; testing of a domestic ozone generator. J. Clin. Pathol. 36, 1102–1104.

Eaks, I.L., Morris, L.I., 1956. Respiration of cucumber fruits associated with physiological injury at chilling temperatures. Plant Physiol. 31 (4), 308–314.

Eckert, J.W., Ratnayake, M., 1983. Host-pathogen interactions in postharvest diseases. In: Lieberman, M. (Ed.), Postharvest Physiology and Crop Preservation. Plenum Press, New York, NY, pp. 247–264.

Edward, D.G., Lidwell, O.M., 1943. Studies on air-borne virus infections. III. The killing of aerial suspensions of influenza virus by hypochlorous acid. J. Hyg. (London) 43 (3), 196–200.

El Nahal, A.K.M., 1953. Responses of pests to fumigation. IV. The responses of Calandra spp. to reduced pressure. Bull. Entomol. Res. 44, 651–656.

El-Goorani, M.A., Sommer, N.F., 1979. Suppression of postharvest plant pathogenic fungi by carbon monoxide. Phytopathology 69, 834.

Elford, W.J., Van den Ende, J., 1941. An investigation of the merits of ozone as an aerial disinfectant. J. Hyg. (Cambridge) 42, 240–265.

Fathalah, K.A., Elsayed, M.M., Taha, I.S., Sabbagh, J., 1986. Numerical study evaporation–condensation in a vertical diffusion gap. Heat Mass Transf. 20 (4), 301–309.

Feller, U., 2006. Stomatal opening at elevated temperature: an underestimated regulatory mechanism. Gen. Appl. Plant Physiol. 19–31 (Special Issue 2006).

Fernández-Maculet, J.C., Dong, J.G., Yang, S.F., 1993. Activation of 1-aminocyclopropane-1-carboxylate oxidase by carbon dioxide. Biochem. Biophys. Res. Comm. 193, 1168–1173.

Fidler, J.C., 1951. A comparison of aerobic and anaerobic respiration of apples. J. Exp. Bot. 2, 41.

Fleurat-Lessard, P., 1981. Ultrastructural features of the starch sheath of the primary pulvinus after gravistimulation of the sensitive plant (*Mimosa pudica* L.). Protoplasma 105, 177–184.

Fleurat-Lessard, F., Le Torc'h, J.-M., Marchegay, G., 1990. Effect of temperature on insecticidal efficiency of hypercarbic atmospheres against three insect species of packaged food stuff, Proceedings of Seventh International Working Conference on Stored-Product Protection, vol. 1, pp. 676–684.

Follet, P.A., 2010. Irradiation to control insects in exported fresh commodities: pioneering generic doses. Trans. Am. Nucl. Soc. 102, 17–18.

Follet, P.A., Armstrong, J.W., 2004. Revised irradiation doses to control Melon fly, Mediterranean fruit fly, and Orienta fruit fly (Diptera: Tephritidae). J. Econ. Entomol. 97 (4), 1254–1262.

Follstad, M.N., 1966. Mycelial growth rate and sporulation of *Alternaria tenuis, Botrytis cinerea, Cladiosporium herbarum,* and *Rhizopus stolonifer* in low oxygen and high carbon dioxide atmospheres. Phytopathology 56, 1098–1099.

Foote, C.S., Groyne, T.E., Lehrer, R.I., 1983. Assessment of chlorination in human neutrophils. Nature 301, 715–716.

Frazier, C.W., Westhoff, C.D., 1988. Food Microbiology, fourth ed. McGraw-Hill, New York, NY.

Frenkel, C., Jen, J.J., 1989. In: Eskin, T.N.A.M. (Ed.), Quality and Preservation of Vegetables. CRC Press, Boca Raton, Florida, USA, pp. 53–74, Chapter 2.

Frenkel, C., Patterson, M.E., 1969. The effect of carbon dioxide on succinic dehydrogenase in pears during cold storage. Hort. Sci. 4, 165.

Fu, J., 2010. The application of new physical preservation methods on chilled meats. Meat Res. 2010-11.

Fuller, E.N., Shettler, P.D., Giddings, J.C., 1966. New method for prediction of binary gas-phase diffusion coefficients. Ind. Eng. Chem. 589 (5), 18–27.

Gac, A., 1956. Contribution to the study of the influence of the relative humidity and the rate of circulation of air on the behavior of harvested fruit. Revue Generale du Froid 33, 363–379, 365–379; 505–533; 733–744; 833; 843; 969–976.

Gao, H., Song, L., Zhou, Y., Yang, Y., Chen, W., Chen, H., et al., 2008. Effects of hypobaric storage on quality and flesh leatheriness of cold-stored loquat fruit. Trans. CSAE 24 (6), 245–249.

Gao, H.Y., Chen, H.J., Chen, W.X., Yang, J.T., Song, L.L., Jiang, Y.M., et al., 2006. Effect of hypobaric storage on physiological and quality attributes of loquat fruit at low temperature. Acta Hortic. 712, 269–274.

Geweely, N.S.I., Nawar, L.S., 2006. Sensitivity to gamma irradiation of post-harvest pathogens of pear. Int. J. Agric. Biol. 8 (6), 710–716.

Ghabrial, A., Luschnig, S., Metzstein, M.M., Krasnow, M.A., 2003. Branching morphogenesis of the *Drosophila* tracheal system. Annu. Rev. Cell Dev. Biol. 19, 623–647.

Gill, C.O., Harrison, J.C.L., 1989. The storage life of chilled pork packaged under carbon dioxide. Meat Sci. 26, 313–324.

Gillooly, J.F., Brown, J.H., West, G.B., Savage, V.M., Charnov., E.L., 2001. Effects of size and temperature on metabolic rate. Science 293, 2248–2251.

Girsch, G.K., 1978. Susceptibility of full grown larvae of *Trogoderma granarium* Everts to varying concentrations of carbon dioxide at low oxygen tension. Bull. Grain Technol. 16, 199.

Goddard, D.R., 1947. The respiration of cells and tissues. In: Höber, R. (Ed.), Physical Chemistry of Cells and Tissues. Churchill, London, pp. 371–444.

Goldschmidt, E.E., Huberman, M., Goren, R., 1993. Probing the roles of endogenous ethylene in the de-greening of citrus fruits with ethylene antagonists. Plant Growth Regul. 12, 325–329.

Goliás, J., Kobza, F., 2002. Ethanol content in cut roses at low oxygen atmosphere storage. Hort. Sci. (Prague) 29 (4), 148–152.

Goos, R.D., Tschirsch, M., 1962. Effect of environmental factors on spore germination, spore survival and growth of *Gleosporium musarum*. Mycologia 34, 353.

Gray, J., 2005. Guard cells: transcription factors regulate stomatal movements. Curr. Biol. 15 (15), R593–R595.

Greer, B.W., Heinstra, P.W., McKechnie, S.W., 1998. The biological basis of ethanol tolerance in *Drosophila*. Comp. Biochem. Physiol. (B) 105, 201–229.

Gundel, L.A., Sullivan, D.P., Katsapov, G.Y., Fisk, W.J., 2002. A Pilot Study of Energy Efficient Air Cleaning for Ozone. Indoor Environment Department, Environmental Energy Technologies Division, Lawrence Berkeley National Laboratory, Berkeley, CA, pp. 94720.

Haard, N.F., Lee, Y.Z., 1982. Hypobaric storage of Atlantic salmon in a carbon dioxide atmosphere. Can. Inst. Food Sci. Technol. J. 15, 68–71.

Haard, N.F., Martins, I., Newberry, T., Bota, R., 1979. Hypobaric storage of Atlantic herring and cod. Can. Inst. Food Sci. Technol. 12, 84–87.

Hacohen, N., Kramer, S., Sutherland, D., Hiromi, Y., Krasnow, M.A., 1998. Sprouty encodes a novel antagonist of FGF signaling that patterns apical branching of the *Drosophila* airways. Cell 92 (2), 253–263.

Hadley, N.F., 1994. Ventilatory patterns of respiratory transpiration in adult terrestrial insects. Physiol. Zool. 67, 75–189.

Hagenmaier, D.R., Baker, R.A., 1997. Low-dose irradiation of cut iceberg lettuce in modified atmosphere packaging. J. Agric. Food Chem. 45, 2864–2868.

Hall, D.O., Rao, K.K., 1999. Photosynthesis. Cambridge University Press, Cambridge, UK.

Hamlin, C., Champ, Y.S., 1974. Optimal condition for mutagenesis by ozone in *E. coli* K12. Mutat. Res. 24, 271–279.

Hammond, D.G., Kubo, I., 2001. Volatile aldehydes as pest control agents. US Patent 6201026.

Han, J.-q., Zhang, Y.-l., 2006. Research on technology of hypobaric storage of stem of garlic scape. J. Jilin Agric. Univ. 2006-02.

Han, j.-q., Zhang, Y.-l., Shi, X.-x., 2006. Study on fresh-keeping technology of hypobaric and ozone on Dong jujube. J. Northwest Sci-Tech. Univ. Agric. For. (Nat. Sci. Ed.) 2006-11.

Hao, X.-l., Wang, R.F., 2004. The physiological changes in Lizao under hypobaric conditions. Food Sci. Tech. 7, 91–93 (in Chinese).

Hardenburg, R.E., Watada, A.E., Wang, C.L., 1986. The Commercial Storage of Fruits and Vegetables, and Florist and Nursery Stocks, Handbook No. 66 (revised). USDA.

Harding, P.R., 1968. Effect of ozone on *Penicillium* mold decay and sporulation. Plant Dis. Rep. 52, 245–247.

Harrison, J.E., Schultz, J., 1976. Studies on the chlorinating activity of myeloperoxidase. J. Biol. Chem. 251, 1371–1374.

Hartman, R.E., Keen, N.T., Long, M., 1972. Carbon dioxide fixation by *Verticillium albo-atrum*. J. Gen. Microbiol. 73, 29–34.

Hashmi, M.S., East, A.R., Palmer, J.S., Heyes, J.A., 2013. Hypobaric treatment stimulates defence-related enzymes in strawberry. Postharvest Biol. Technol. 85, 77–82.

Hatton, T.T., Spalding, D.H., 1990. Controlled atmosphere storage of some tropical fruits. In: Calderon, M., Barkai-Golan, R. (Eds.), Food Preservation by Modified Atmospheres. CRC Press, Boca Raton, FL, pp. 303–315.

Hatton, T.T., Pantastico, E.B., Akamine, E.K., 1975. Individual commodity requirements. In: Haard, N.F., Salunkhe, D.K. (Eds.), Symposium: Postharvest Biology and Handling of Fruits and Vegetables. AVI Publishing Co., Westport, CT, pp. 201–218.

Hazelhoff, E.H., 1927. Die regerlierung der atmung bei insekten and spinnen. Z. Vergleich. Physiol. 5, 179–190.

He, C.J., Davies, F.T., Lacey, R.E., Drew, M.C., Brown, D.L., 2003. Effect of hypobaric conditions on ethylene evolution and growth of lettuce and wheat. J. Plant Physiol. 160 (11), 1341–1350.

Heberlein, U., 2000. Genetics of alcohol-induced behaviors in *Drosophila*. Alcohol Res. Health 24, 185–188.

Henry, J.R., Harrison, J.F., 2004. Plastic and evolved responses of larval tracheae and mass to varying atmospheric oxygen content in *Drosophila melanogaster*. J. Exp. Biol. 207, 3559–3567.

Herreid, C.F., 1980. Hypoxia in invertebrates. Comp. Biochem. Physiol. A67, 311.

Holzwarth, G., Balmer, R.G., Soni, L., 1984. The fate of chlorine and chloramines in cooling towers. Henry's Law constants for flash-off. Water Res. 18, 1421–1427.

Hu, X., Zhang, C.-f., Zheng, X.-z., 2012. A new model investigation on low pressure cold chain for fresh agricultural product. J. Changjiang Vegetables. 2009-08.

Huang, S., Zhang, J.-s., 2001. Effect of low ethylene and low pressure treatment on ethylene biosynthesis by persimmon fruit. Acta Bot. Boreali-Occidentalia Sin. 2001-02.

Huang, S., Zhang, J.-s., Li, W.-p., 2003. Effect of hypobaric treatment on the softening physiology in post-harvest persimmon fruits. J. Northwest Sci-Tech. Univ. Agric. For. 2003-05.

Hughes, P.A., Thompson, A.K., Plumbley, R.A., Seymour, G.B., 1981. Storage of Capsicums (*Capsicum-annum* cultivar Bellboy) under controlled atmosphere, modified atmosphere and hypobaric conditions. J. Hortic. Sci. 56 (3), 261–266.

Hulme, A.C., 1956. Carbon dioxide injury and the presence of succinic acid in apples. Nature 178, 218–219.

Husband, P.M., 1982. The Grumman Dormavac container: its potential for the storage and transport of chilled lamb. Meat Research Report, Division of Food Research, Commonwealth Scientific and Industrial Research Organization, Australia. No 1/82, pp. 1–16.

Ilangantileke, S.O., Turla, L.O., Chen, R.C., 1989. Pretreatment and Hypobaric Storage for Increased Storage Life of Mangos. Book published by American Society of Agricultural Sciences, St. Joseph, Mich. 15 pps.

Imahori, Y., Uemura, K., Kishioka, I., Fujiwara, H., Tilio, A.Z., Ueda, Y., et al., 2005. Relationship between low-oxygen injury and ethanol metabolism in various fruits and vegetables. Acta Hortic. 682: Proceedings of the Fifth International Postharvest Symposium. Verona, Italy, June 6–11, 2004.

Imms, A.D., 1949. Outlines of Entomology. Methuen & Co. Ltd, London.

Imoolehin, E.D., Grogan, R.G., 1980. Effects of oxygen, carbon dioxide and ethylene on growth, sclerotial production, germination, and infection by *Sclerotinia minor*. Phytopathology 70, 1158.

Izumi, H., Nakatani, T., Ogikuba, H., 1999. Controlled-atmosphere storage of fresh-cut spinach at various temperatures. HortScience 34 (3), 507.

Jadhav, S.J., Patil, B.C., Salunkhe, D.K., 1973. Control of potato greening under hypobaric storage. Food Eng. 45 (8), 111.

Jamieson, L.E., Wimalaratne, S.K., Bycroft, B.L., van Epenhuijse, C.W., Page, B.B.C., Somerfield, K.G., et al. 2009. Feasibility of ozone for treating sea containers. MAF Biosecurity New Zealand Technical Paper No. 2010/01 (December).

Jamieson, W., 1980a. Dormavac low-pressure storage of perishable commodities. Prog. Food Nutr. Sci. 4, 61–72.

Jamieson, W., 1980b. Use of hypobaric conditions for refrigerated storage of meat, fruits and vegetables. Food Technol. March, 64–71.

Jarecki, J., Johnson, E., Krasnow, M.A., 1999. Oxygen regulation of airway branching in *Drosophila* is mediated by branchless FGF. Cell 99, 211–220.

Jay, E.G., Cuff, W., 1981. Weight loss and mortality of three life stages of *Tribolium castaneum* (Herbst) when exposed to four modified atmospheres. J. Stored Prod. Res. 17, 117–124.

Jay, E.G., Pearman, G.C., 1971. Susceptibility of two species of *Tribolium* (Coleoptera: Tenebrionidae) to alterations of atmospheric gas concentrations. J. Stored Prod. Res. 7, 181–186.

Jay, E.G., Arbogast, R.T., Pearman, G.C., 1971. Relative humidity: its importance in the control of stored-product insects with modified atmospheric gas concentrations. J. Stored Prod. Res. 6, 325.

Jiang, W.-l., Wang, S.-q., Meng, J., Song, Q.-w., 2009. Effect of hypobaric storage on preservation of cucumbers. Storage Process. 2009-04.

Jiao, S., Johnson, J.A., Fellman, J.K., Mattinson, D.S., Tang, J., Davenport, T.L., et al., 2012. Evaluating the storage environment in hypobaric chambers used for disinfecting fresh fruits. Biosyst. Eng. 111 (3), 271–299.

Jõgar, K., Kuusik, A., Metspalu, L., Külli, H., Luik, A., Mänd, M., et al., 2004. The relation between the patterns of gas exchange and water loss in diapausing pupa of the white butterfly *Pieris brassicae* (Lepidoptera: Pieridae). Eur. J. Entomol. 101, 467–492.

Johnson, B.Y., 1974. Chilled vacuum-packed beef. CSRIO Food Res. Quart. 34, 14.

Kader, A.A., 1975. Hypobaric storage of horticultural commodities. Postharvest Physiology Seminar. University of California, Davis, CA, June 5, 1975.

Kader, A.A., 1986. Potential applications of ionizing radiation in postharvest handing of fresh fruits and vegetables. Food Tech. 40 (6), 117–121.

Kader, A.A., 2002. Chapter 4: Postharvest technology of horticultural crops. In: Kader, A.A. (Ed.), Postharvest Biology and Technology, third ed. United California Agricultural and Natural Resources (ANR) Publication 3311, pp. 39–47.

Kader, A.A., 2005. Increasing food availability by reducing post harvest losses. Acta Hortic. 682, 21649–22175.

Kader, A.A., Rolle, R.S., 2004. The role of postharvest management in assuring the quality and safety of horticultural produce. FAO Agric. Serv. Bull., 152.

Kang, M.-l., Zhang, P., 2001. Research progress in the theory and technology of hypobaric storage. Food Mach. 82 (2), 9–10 (in Chinese).

Kao, C.H., Yang, S.F., 1982. Light inhibition of the conversion of ACC to ethylene in leaves is mediated through carbon dioxide. Planta 155, 261–266.

Kao, N.Y., 1971. Extension of storage life of bananas by gamma radiation. Disinfestation of fruit by irradiation: Proceedings of a Panel, Honolulu, Accession No. 116825, Report No: AGE-PL/422/13; Fiche No. 16825 (STI/PUB/299), pp. 125–136.

Ke, D., Kader, A., 1989. Tolerance and responses of fresh fruits to oxygen levels at or below 1%. In: Fellman, J.K. (Ed.), Proceedings of the Fifth International CA Research Conference, Wenatchee, WA, vol. 2—other commodities and storage recommendations, pp. 209–216.

Ke, D., Yahia, E., Hess, B., Zhou, L., Kader, A.A., 1995. Regulation of fermentative metabolism in avocado fruit under oxygen and carbon dioxide stresses. J. Am. Soc. Hort. Sci. 120 (3), 481–490.

Ke, D.Y., Kader, A.A., 1992. Potential of controlled atmospheres for postharvest insect disinfestation of fruits and vegetables. Postharvest News Inf. 3, 31N–37N.

Ke, D.Y., Mateos, M., Kader, A.A., 1993. Regulation of fermentative metabolism in fruits and vegetables by controlled atmospheres. In: Proceedings of the Sixth International CA Research Conference, Ithaca, NY, vol. 1, pp. 63–77.

Kestler, P., 1985. Respiration and respiratory water loss. In: Hoffmann, K.H. (Ed.), In: Environmental Physiology and Biochemistry of Insects. Springer-Verlag, Heidelberg, Berlin, pp. 137–183.

Kettle, A.J., Winterbourn, C.C., 1997. Myeloperoxidase: a key regulator of neutrophil oxidant production. Redox Rep. 3, 3–15.

Kidd, F., West, C., 1927a. Gas storage of fruit. Great Britain Department of Scientific Industrial Research. Food Investigation Board Report. vol. 30, 87.

Kidd, F., West, C., 1927b. A relation between the concentration of O_2 and CO_2 in the atmosphere, rate of respiration and the length of storage of apples. Food Invest. Board Rep. London for 1925, 1926, 41–42.

Kidd, F., West, C., 1945. Respiratory activity and duration of life of apples gathered at different stages of development and subsequently maintained at a constant temperature. Plant Physiol. 20, 467–504.

Kilonzo-Nthenge, A.K., 2012. Gamma irradiation of fresh produce. <http://www.intechopen.com/books/gamma-radiation>.

Kirk, H.G., Andersen, A.S., 1986. Influence of low-pressure storage on stomatal opening and rooting of cuttings. Acta Hortic. 181, 393–397.

Kirk, H.G., Andersen, A.S., Veierskov, B., Johansen, E., Aabrandt, Z., 1986. Low-pressure storage of hibiscus cuttings. Effect on stomatal opening and rooting. Ann. Bot. 58 (3), 389–396.

Knee, M., Hatfield, S.G.S., 1981. The metabolism of alcohols by apple tissues. J. Sci. Food Agric. 32 (6), 593–600.

Knox, W.E., Stumpf, P.K., Green, D.E., Auerbach, V.H., 1948. The inhibition of sulfhydryl enzymes as the basis of the bacterial action of chlorine. J. Bacteriol. 55 (4), 451–458.

Knudsen, J.G., Bell, K.J., Holt, A.D., Hottel, H.C., Sarofim, A.F., Standiford, F.C., et al., 1984. Heat transmission. In: Crawford, H.B., Eckes, B.E. (Eds.), Perry's Chemical Engineer's' Handbook. McGraw-Hill, New York, NY, pp. 1–68 (Section 10).

Koyama, S., Yashuhara, K., Yara, T., 2002. Study on mist formation from humid air cooled in a rectangular tube. Eng. Sci. Rep. Kyushu Univ. 24 (2), 187–193.

Krafsur, E.S., Graham, C.L., 1970. Spiracular responses of *Aedes* mosquitos to carbon dioxide and oxygen. Ann. Entomol. Soc. Am. 63 (3), 691–696.

Krebs, H.A., 1943. Carbon dioxide assimilation in heterotrophic organisms. Annu. Rev. Biochem. 12, 529–550.

Kreith, F., Bohn, M.S., 1997. Principles of Heat Transfer. PWS Publishing Co., Boston.

Krogh, A., 1920. Diffusion theory of tracheal respiration. Arch. Ges. Physiol. 179, 95–120.

Krogh, A., 1941. The Comparative Physiology of Respiratory Mechanisms. University of Penn Press, Philadelphia, pp. viii–172.

Kubo, Y., Inaba, A., Nakamura, R., 1996. Extinction point and critical oxygen concentration in various fruits and vegetables. J. Jpn. Soc. Hort. Sci. 65 (2), 397–402.

Kui, H.Y., 1971. Extension of storage life of bananas by gamma irradiation. In: From Proceedings of a Panel, Honolulu, HI. December 7–11, 1970, pp.125–136.

Kuts, P.S., Pikus, J.F., Kalinina, L.S., 1975. Coefficient of internal mass transfer in electrical-grade cellulosics under vacuum and under atmospheric pressure. J. Eng. Phys. Thermophys. 26 (4), 447–452.

LaRue, J.H., Johnson, R.S., 1989. Peaches, Plums and Nectarines. Growing and Handling for Fresh Markets. United California Division of Agricultural and Natural Resources (Publication 3331).

Ladaniya, M.S., Singh, S., Wadhawan, A.K., 2003. Response of "Nagpur" mandarin, "Mosambi" sweet orange and "kagzi" acid lime to gamma radiation. Radiat. Phys. Chem. 67, 665–675.

Lagunas-Solar, M.C., Essert, T.K., Zeng, N.X., Pina-U, C., Truong, T.D., 2003. Non-thermal metabolic stress disinfestation and disinfection method for fresh agricultural products. In: Proceedings of the 2003. International Research Conference on Methyl Bromide Alternatives and Emissions Reductions, 75-1 to 75-4, November 3–6, 2003, San Diego, CA.

Lagunas-Solar, M.C., Essert, T.K., Pina-U, C., Xeng, N.X., Truong, T.D., 2006. Metabolic stress disinfestation (MSDD): a new, non-thermal, residue-free process for fresh agricultural products. J. Sci. Food Agric. 86, 1814–1825.

Laohakunjit, N., Uthairatakij, A., Kerdchoechuen, O., Chatpaisarn, A., Photochanachai, S., 2005. Identification of changes in volatile compound in γ-irradiated mango during storage. International Symposium "New Frontier of Irradiated Food and Non-food Products. Sept. 22–23, 2005. KMUTT, Bangkok, Thailand, pp. 8.

Laurin, É., Nunes, N., Émond, J.P., Brecht, J.K., 2003. Quality of strawberries after simulated air freight conditions. International Conference on Quality in Chains. An Integrated View on Fruit and Vegetable Quality, ISHS Acta Horticulturae 604 (2), 659.

Laurin, É., Nunes, M.C.N., Émond, J.-P., Brecht, J.K., 2005. Vapor-pressure deficit and water loss patterns during simulated air shipment and storage of Beit Alpha cucumber. In: Proceedings of the Fifth International Postharvest Symposium. Verona. Italy. ISHS. Acta Hortic. 682: 1697–1704.

Laurin, É., Nunes, M.C.N., Émond, J.-P., Brecht, J.K., 2006. Residual effect of low pressure stress during simulated air transport on Beit Alpha-type cucumbers: stomata behavior. Postharvest Biol. Technol. 41 (2), 121–127.

Lavista-Llanos, S., Centanin, L., Irisarri, M., Russo, D.M., Gleale, J.M., et al., 2002. Control of the hypoxic response in Drosophila melanogaster by the basic helix-loop-helix PAS protein similar. Mol. Cell Biol. 22, 6842–6853.

Lentz, C.P., Rooke, E.A., 1964. Rates of water loss from apples under refrigerated storage conditions. Food Tech. 18, 119–121.

Leshuk, J.A., Saltveit, M.E., 1990. Controlled atmosphere storage requirements and recommendations for vegetables. In: Calderon, M., Barkai-Golan, R. (Eds.), Food Preservation by Modified Atmospheres. CRC Press, Boca Raton, FL/Boston/Ann Arbor, pp. 315–352.

Levine, L.H., Richards, J.T., Wheeler, R.M., 2009. Super-elevated CO_2 interferes with stomatal closure in soybean (*Glycine max*). J. Plant Physiol. 166 (9), 903–913.

Li, H., Chai, X.-S., Deng, Y., Zhan, H., Fu, S., 2009. Rapid determination of ethanol in fermentation headspace gas chromatography. J. Chromatogr. [A] 1216 (1), 169–172.

Li, J., Zhou, Y., 1993. Ethylene metabolism of stored "Quanxing" apricot fruits under hypobaric atmosphere. J. Northwest Sci-Tech. Univ. Agric. For. 1993-02.

Li, J.-K., Zhang, P., Zhang, P., 2010. Effects of hypobaric storage on storage and physiology of Mopan persimmons. Storage Process, 2010-05.

Li, L., Wang, R.-f., 2007. A review of hypobaric storage in fruits and vegetables. J. Shanxi Agric. Sci. March, 2007-03.

Li, W.-x., Zhang, M., 2005. Effects of combined hypobaric and atmospheric cold storage on the preservation of honey peach. Int. Agrophys. 19 (3), 1–6.

Li, W.-x., Zhang, M., 2006. Effect of three-stage hypobaric storage on cell wall components, texture and cell structure of green asparagus. J. Food Eng. 77 (1), 112–118.

Li, W.-x., Zhang, M., Yu, H.-q., 2004. Study on hypobaric storage of green *Asparagus officinalis* L. J. Wuxi Univ. Light Ind. 23 (6), 38–432 (in Chinese).

Li, W.-x., Zhang, M., Tao, F., 2005a. Study of vacuum precooling combined with hypobaric storage on keeping Honey peach fresh. J. Wuxi Univ. Light Ind. 2005-05.

Li, W.-x., Zhang, M., Yu, H.-q., 2005b. Study on hypobaric storage of green asparagus. J. Food Eng. 73 (3), 225–230.

Li, W.-x., Zhang, M., Yu, H.-q., 2006. Study on hypobaric storage of green asparagus. J. Food Eng. 73, 225–230.

Li, W.-x., Zhang, M., Wang, S.-j., 2007a. Effect of three-stage hypobaric storage on membrane lipid peroxidation and activities of defense enzyme in green asparagus. Food Sci. Tech. 41, 2175–2181.

Li, W.-x., Li, W.-x., Sun, B.-s., Wang, J.-f., Zhang, M., 2007b. Mathematical evaluation of three-stage hypobaric treatment on the three-stage hypobaric storage technology of green asparagus. Ludong Univ. J. (Nat. Sci. Ed.) 2007-02.

Li, W.-x., Qiu, H.-w., Sun, P., Zhang, M., 2008. Effects of three-stage hypobaric storage on green asparagus physiologic changes. Food Sci. 2008-02.

Liang, Y.K., Dubos, C., Dodd, I.C., Holroyd, G.H., Hetherington, A.M., Campbell, M.M., 2005. AtMYB61, an R2R3-MYB transcription factor controlling stomatal aperture in *Arabidopsis thaliana*. Curr. Biol. 15 (13), 1201–1206.

Lighton, J.R.B., 1994. Discontinuous ventilation in terrestrial insects. Physiol. Zool. 67, 142–162.

Lighton, J.R.B., Berrigan, D., 1995. Questioning paradigms: caste-specific ventilation in harvester ants, *Messor pergandei* and *M. julianus* (Hymenoptera: Formicidae). J. Insect Physiol. 39, 687–699.

Lighton, J.R.B., Garrigan, D.A., Duncan, F.D., Johnson, R.A., 1993. Spiracular control of respiratory water loss in female alates of the harvester ant *Pogonomyrmex rugosus*. J. Exp. Biol. 179, 233–244.

Liley, P.E., Reid, R.C., Buck, E., 1984. Physical and chemical data. In: Crawford, H.B., Eckes, B.E. (Eds.), Perry's Chemical Engineers' Handbook. McGraw-Hill, New York, NY, pp. 1–291. Section 3.

Lipton, W.J., 1975. Controlled atmospheres for fresh vegetables and fruits—why and when. In: Haard, N.F., Salunkhe, D.K. (Eds.), Symposium: Postharvest Biology and Handling of Fruits and Vegetables. AVI Publishing Co., Westport, CT, pp. 130–142.

Lipton, W.J., Harvey, J.M., Couey, H.M., 1967. Conclusions about radiation. United Fresh Fruit and Vegetable Association Yearbook. 1967, Alexandria, VA, pp. 173.

Lister, B.W., 1952. The decomposition of hypochlorous acid. Can. J. Chem. 30, 879−889.

Liu, H., Jiang, Y., 2008. Experimental study on hypobaric storage of Chinese gooseberry. Dhunese Assoc. Refrig. [C] 2008.

Liu, Y.-B., 2003. Effects of controlled atmosphere treatments on insect mortality and lettuce quality. J. Econ. Entom. 96 (4), 1100−1107.

Locke, M., 1959. The co-ordination of growth in the tracheal system of insects. Quart. J. Microsc. Sci. 99, 373−391.

Lockhardt, C.L., 1967. Influence of controlled atmospheres on the growth of *Gloeosporium album in vitro*. Can. J. Plant Sci. 47, 649−651.

Lockhardt, C.L., 1968. Influence of various carbon dioxide and oxygen concentrations on the growth of *Fusarium oxysporum in vitro*. Can J. Plant Sci. 48, 451.

Lockhardt, C.L., 1969. Effect of CA on storage rot pathogens. In: Dewey, D.H., Herner, R.C., Dilley, D.R. (Eds.), Controlled Atmospheres for the Storage and Transport of Horticultural Crops. Proceedings of the National CA Research Conference at Michigan State University, January 27 and 28, pp. 113−121.

Loudon, C., 1989. Tracheal hypertrophy in mealworms: design and plasticity in oxygen supply systems. J. Exp. Biol. 147, 217−235.

Lougheed, E.C., Murr, D.P., Berard, L., 1977. LPS—Great expectations. In: Dewey, D.H. (Ed.), Horticultural Report: Controlled Atmospheres for the Storage and Transport of Perishable Agricultural Commodities. Proceedings of the Second National CA Research Conference, April 5−7, at Michigan State University, pp. 3−44.

Lougheed, E.C., Murr, D.P., Bérard, L., 1978. Low-pressure storage of horticultural crops. HortScience 13 (1), 21−27.

Louguet, P., 1968. Influence de l'anaerobios sur le mouvement des stomata á l'obscurité chez le *Pelargoneum* x *hortorum*. Physiol. Veg. 6, 83−92.

Louguet, P., 1972. Influence de la pression partielle de oxygéne sur la vitesse d'overture et de fermenture des stomates de *Pelargonium* x *hortorum* á l'obscutité. Physiol. Veg. 6, 8392.

Luoyang, L., 2009. The application research development of hypobaric storage in fruits and vegetables. Acad. Period. Farm Prod. Process. 2009-04.

Lutz, J.M., Hardenburg, R.E., 1968. The Commercial Storage of Fruits, Vegetables, and Florist and Nursery Stocks, Agric. Handbook No. 66. USDA.

Machin, J., Kestler, P., Lampert, G.J., 1991. Simultaneous measurement of spiracular and cuticular water losses in *Periplaneta americana*: implications for whole-animal mass loss studies. J. Exp. Biol. 161, 439−453.

Maity, J.P., Kar, S., Banerjee, S., Sudershan, M., Chakraborty, A., Shantrs, S.C., 2011. Effects of gamma radiation on fungal infected rise (*in vitro*). Int. J. Radiat. Biol. 87 (11), 1097−1103.

Masterman, A.T., 1938. Air purification in inhabited rooms by spraying or atomizing hypochlorites. J. Ind. Hyg. 20, 278−288.

Masterman, A.T., 1941. Air purification by hypochlorous acid gas. J. Ind. Hyg. 41 (1), 44−64.

Mattheis, J.P., Buchanan, D.A., Fllman, J.K., 1991. Change in apple fruit volatiles after storage in atmospheres inducing anaerobic metabolism. J. Agric. Food Chem. 39, 1602−1605.

Maurel, C., Santoni, V., Luu, D.-T., Wudick, M.M., Verdoucq, L., 2009. The cellular dynamics of plant aquaporin expression and function. Plant Biol. 12, 690−698.

Maxie, E.C., Abdel-Kader, A.S., 1966. Food irradiation – Physiology of fruits as related to feasibility of the technology. Adv. Food Res. 15, 105.

McKeown, A.W., Lougheed, E.C., 1980. Low-pressure storage of some vegetables. Acta Hortic. 116, 83–100.

McRae, D.G., Coker, J.A., Legge, R.L., Thompson, J.E., 1983. Bicarbonate/CO_2-facilitated conversion of 1-aminocyclopropane-1-carboxylic acid to ethylene in a model system and intact tissues. Plant Physiol. 73, 784–790.

Mellanby, K., 1934. The site of loss of water from insects. Proc. R. Entomol. Soc. London (B) 116, 130–149.

Mermelstein, N.H., 1979. Hypobaric transport and storage of fresh meats and produce earns IFT Food Technology Industrial Achievement Award. Food Tech. 33, 32–35, 38-40.

Metzger, R.J., Krasnow, M.A., 1999. Genetic control of branching morphogenesis. Science 284 (5420), 1635–1639.

Milan, N.F., Kacsoh, B.Z., Schlenke, T.A., 2012. Alcohol consumption as a self-medication against blood-borne parasites in the fruit fly. Curr. Biol. 22 (6), 488–493.

Mill, P.J., 1985. Structure and physiology of the respiratory system. In: Kerkut, G.A., Gilbert, L.I. (Eds.), Comprehensive Insect Physiology, Biochemistry and Pharmacology, vol. 3. Pergamon Press, Oxford, pp. 517–593.

Miller, P.L., 1973. Spatial and temporal changes in the coupling of cockroach spiracles to ventilation. J. Exp. Biol. 59, 137–148.

Miller, P.L., 1981. Ventilation in active and inactive insects. In: Herreid, C.F., Fourtner, C.R. (Eds.), Locomotion and Energetics in Arthropods. Plenum Press, New York, NY, pp. 367–390.

Minas, I.S., Karaoglanidis, G.S., Manganaris, G.A., Vasilakakis, M., 2011. Gaseous ozone treatment of kiwifruit during cold storage induces resistance to stem-end rot caused by the fungal pathogen Botrytis cinerea. Int. Cong. Postharvest Pathol., Lleida (April 11–14).

Mingli, K., 2001. Research progress in the theory and technology of hypobaric storage. Food Mach. 82 (2), 9–10.

Mitcham, E., Martin, T., Zhou, S., 2006. The mode of action of insecticidal controlled atmospheres. Bull. Entomol. Res. 96, 213–222.

Mitchell, D.J., Zentmyer, G.A., 1971. Effects of oxygen and carbon dioxide tensions on growth of several species of Phytophthora. Phytopathology 61 (2), 787–791.

Monod, J., 1942. La Croissance des Cultures Bactériennes, 2ll pps. Hermann et Coe, Paris.

Moreno, M.A., CastellPerez, M.E., Gomes, C., Da Silva, P.F., Kim, J., Moreira, R.G., 2007. Optimizing electron beam irradiation of "Tommy Atkins" mangos (Mangifera indica L.). J. Food Process Eng. 30, 436–457.

Mortimer, N.T., Moberg, K.H., 2009. Regulation of Drosophila embryonic tracheogenesis by dVHL and hypoxia. Dev. Biol. 329 (2), 294–305.

Mudd, S., 1944. Current progress in sterilization of air. Br. Med. J. (London) July (15), 67–80.

Munro, H.A.U., 1959. The response of Tenebroides mauritanicus (L.) and Tenebrio molitor (L.) to methyl bromide at reduced pressures. J. Sci. Food Agric. 7, 366–379.

Murata, T., Ueda, Y., 1967. Studies on the storage of spinach. J. Jpn. Soc. Hort. Sci. 36, 449.

Murray, D.R., 1997. Carbon Dioxide and Plant Responses. John Wiley & Sons, New York, NY.

Murray, P.W., 2011. Microbiology and Immunology. Sixth ed. Chapter 3. Mosby, Inc.

Nadas, A., Olmo, M., Garcia, J.M., 2003. Growth of Botrytis cinerea and strawberry quality in ozone enriched atmospheres. J. Food Sci. 68, 1798–1802.

Narayanan, E.S., Bhambhani, H.J., 1956. Effect of reduced pressures on *Tribolium castaneum* Hbst. (Tenebrionidae, Coleoptera) and *Trogoderma granarium* Everts (Dermestidae, Coleoptera). Indian J. Entomol. 18, 196–198.

National Academy of Sciences, 1978. Postharvest Food Losses in Developing Countries. NAS, Washington, DC.

Navarro, S., 1974. Studies on the effect of alterations in pressure and composition of atmospheric gases on the tropical warehouse moth, *Ephestia cautella* (Wlk.), as a model for stored-product insects (Ph. D. thesis), Senate of Hebrew University, Jerusalem, pp. 118 (in Hebrew with English summary).

Navarro, S., 1975. Effect of oxygen concentrations on *Tribolium castaneum* (Herbst) adults exposed to different relative humidities. Rep. Stored Prod. Res. Lab 13–21 (in Hebrew with English summary).

Navarro, S., 1978. The effects of low oxygen tensions on three stored-product insect pests. Phytoparasitica 6, 51.

Navarro, S., Calderon, M., 1978. Mode of action of low atmospheric pressure on *Ephestia cautella* (Wlk.) pupae. Cell. Mol. Life Sci. 35 (5), 620–621.

Navarro, S., Calderon, M., 1979. Mode of action of low atmospheric pressures on *Ephestia cautella* (Wlk.) pupae. Experientia 35, 620–621.

Navarro, S., Donahaye, E., 1972. An apparatus for studying the effects of controlled low pressures and compositions of atmospheric gases on insects. J. Stored Prod. Res. 8, 223–226.

Neil, T.A., Leonard, R.T., Macnish, A.J., 2006. Taking the mystery out of flower care solutions. <www.floristsreview.com/main/June2006/featurestory.html>.

Ng, H., 1969. Effect of decreasing growth temperature on cell yield of *Escherichia coli*. J. Bacteriol. April, 232–237.

Niblock, J.S., 2012. Vivafresh claims world record for fresh-cut flower storage: 60 days. Produce News June 1, 2012.

Noble, P.S., 1991. Physiochemical and Environmental Plant Physiology. Academic Press, San Diego, CA.

Ofuya, T.I., Reichmuth, C., 2002. Effect of relative humidity on the susceptibility of *Callosobruchus maculatus* (Fabricius) (Coleoptera: Bruchidae) to two modified atmospheres. J. Stored Prod. Res. 38, 139–146.

Oxley, T.A., Wickenden, G., 1963. The effect of restricted air supply on some insects which infest grain. Ann. Appl. Biol. 51 (2), 313–324.

Özisik, M.N., 1985. Heat Transfer. A Basic Approach. McGraw-Hill, New York, NY.

Palou, L., Smilanick, J.L., Crisoto, C.H., Mansour, M., 2001. Effect of gaseous ozone on the development of green and blue molds on cold stored citrus fruits. Plant Dis. 85, 632–638.

Palou, L., Crisoto, C.H., Mansour, M., Plaza, P., 2002. Effects of continuous 0.3-ppm ozone exposure on decay development and physiological responses of peaches and table grapes in cold storage. Postharvest Biol. Technol. 24, 39–48.

Palou, L., Smilanick, J.L., Crisoto, C.H., Mansour, M., Plaza, P., 2003. Ozone gas penetration and control of sporulation of *Penicillium digitatum* and *Penicillium italicum* within commercial packages of oranges during cold storage. Crop Prot. 22, 1131–1134.

Palta, J.P., Stadelmann, E.J., 1980. On simultaneous transport of water and solute through plant cell membranes: evidence for the absence of solvent drag effect and insensitivity of the reflection coefficient. Physiol. Plantarum 50 (1), 83–90.

Pappenheimer, A.M., Hottle, G.A., 1940. Effect of certain purines and CO_2 on growth of strain of group A haemolytic *Streptococcus*. Proc. Soc. Exp. Biol. Med. 44, 645.

Parsons, C.S., Spalding, D.H., 1971. Influence of controlled atmospheres, storage temperatures, and state of ripeness on the development of four fungal rots in tomatoes. HortScience 6 (3), 294.

Parsons, C.S., Anderson, R.E., Penny, R.W., 1970. Storage of mature green tomatoes in controlled atmospheres. J. Am. Soc. Hort. Sci. 95, 791–793.

Paster, N., 1990. Modified atmospheres for preventing molds and mucotoxin production in stored grain. In: Calderon, M., Barkai-Golan, R. (Eds.), Food Preservation by Modified Atmospheres. CRC Press, Boca Raton, FL, pp. 39–55.

Paul, A.L., Schuerger, A.C., Popp, M.P., Richards, J.T., Manak, M.S., Ferl, R.J., 2004. Hypobaric biology: *Arabidopsis* gene expression at low atmospheric pressure. Plant Physiol. 134 (1), 215–223.

Pearman, G.C., Jay, E.G., 1970. The effect of relative humidity on the toxicity of carbon dioxide to *Trilobium casteneum* in peanuts. J. Georgia Entomol. Soc. 5, 61–64.

Pelleu Jr., G.B., Berry, R.F., Holleman, N.G., 1974. Ozone and glycol vapor decontamination of air in a closed room. J. Dental Res. 53 (5), 1132–1137.

Pemadasa, M.A., 1977. Stomatal responses to high temperatures in darkness. Ann. Bot. 41 (5), 969–976.

Perez, A.G., Sanz, C., Rios, J.J., Olias, R., Olias, J.M., 1999. Effects of ozone on postharvest strawberry quality. J. Agric. Food Chem. 47 (4), 1652–1656.

Peterson, J.D., Fowler, P.A., 2004. Leaf temperature of radish (*Raphanus sativus*) in response to increased transpiration under hypobaric conditions. ASGSB Twentieth Annual Meeting, November 9–13, New York (Abstract).

Phillips, T.W., Hulasare, R., Konemann, C., Hallman, G., 2002. Chemical and non-chemical disinfestation of apple maggot flies from apples. <http://mbao.org/2002proc/039phillipst%20mb-alt-abs-phillips-corr.pdf>.

Pimentel, R.M.A., Walder, J.M.M., 2004. Gamma radiation in papaya harvested at three stages of maturity. Scientia Agricola (Piracicaba, Braz.) 61 (2), Mar./Apr., 2004.

Platenius, H., 1942. Effect of temperature on the respiration rate and the respiratory quotient of some vegetables. Plant Physiol. 17, 179–197.

Poneleit, L.S., Dilley, D.R., 1993. Carbon dioxide activation of 1-amino-cyclopropane-1-carboxylate (ACC) oxidase in ethylene biosynthesis. Postharvest Biol. Technol. 3, 191–199.

Price, N.R., Walter, C.M., 1987. A comparison of some effects of phosphine, hydrogen cyanide and anoxia in the lesser grain borer, *Rhyzopertha dominica* (F.) (Coleoptera: Bostrychidae). Comp. Biochem. Physiol. C. 86, 33.

Prince, T.A., 1989. Modified atmosphere packaging of horticultural commodities. In: Brody, A.L. (Ed.), Controlled/Modified Atmosphere/Vacuum Packing of Foods. Food & Nutrition Press, Trumbull, CT, pp. 67–100.

Prosser, C.L., Bishop, D.W., Brown, F.A., Jahn, T.L., Wulff, V.J., 1952. Comparative Animal Physiology. W. B. Saunders Co, Philadelphia and London.

Qu, J.-j., He, Z.-h., Du, Z.-h., Li, F., 2005. Negative ion generation under low pressure and high humidity for fruit's storage. Adv. Tech. Electr. Eng. Energy 2005-02.

Rahn, O., 1941. Notes on the CO_2 requirement of bacteria. Growth 5, 197–199.

Ransom, S.L., 1953. Zymasis and acid metabolism in higher plants. Nature (London) 172, 252.

Ransom, S.L., Walker, D.A., Clarke, I.D., 1957. The inhibition of succinic oxidase by high CO_2 concentrations. Biochem. J. 66, 57.

Ransom, S.L., Walker, D.A., Clarke, I.D., 1960. Effects of carbon dioxide on mitochondrial enzymes from *Ricinus*. Biochem. J. 76, 221–225.

Raschke, K., 1960. Heat transfer between the plant and environment. Ann. Rev. Plant Physiol. 11, 111–126.

Raschke, K., 1972. Saturation kinetics of the velocity of stomatal closing in response to CO_2. Plant Physiol. 49, 229–234.

Raschke, K., 1975. Stomatal action. Annu. Rev. Plant Physiol. 26, 309–340.

Raschke, K., Dunn, W., Lorimer, F., 1970. Saturation kinetic of the velocity of stomatal closing in response to CO_2. Plant Research '70. MSU/AEC Plant Research Laboratory at Michigan State University., pp. 41–43.

Reid, M.S., Jiang, C.Z., 2005. New strategies in transportation for floricultural crops. In: Mencarelli, F., Torutti, P. (Eds.), Proceedings of the Fifth International Postharvest Symposium. Acta Hortic. 682.

Rennie, T.J., Vigneault, C., Raghavan, G.S.V., 2001. Effects of pressure reduction rate on vacuum cooled lettuce quality during storage. Can. Biosyst. Eng. 43, 339–343.

Robinson, J.E., Brown, K.M., Burton, W.G., 1975. Storage characteristics of some vegetables and soft fruits. Ann. Appl. Biol. 81, 339–408.

Rochwell, E., Highberger, J.H., 1927. The necessity of CO_2 for the growth of bacteria, yeasts, and molds. J. Infect. Dis. 40, 438–446.

Rodriguez, J.L., Davies, W.J., 1982. The effects of temperature and ABA on stomata of *Zea mays* L. J. Exp. Bot. 33 (5), 977–987.

Romanazzi, G., Nigro, F., Ippolito, A., Salerno, M., 2001. Effect of short hypobaric treatments on postharvest rots of sweet cherries, strawberries and table grapes. Postharvest Biol. Technol. 22, 1–6.

Romanazzi, G., Nigro, F., Ippolito, A., 2003. Short hypobaric treatments potentiate the effect of chitosan in reducing storage decay of sweet cherries. Postharvest Biol. Technol. 29 (1), 73–80.

Romanazzi, G., Nigro, F., Ippoolito, A., 2008. Effectiveness of a short hyperbaric treatment to control postharvest decay of sweet cherries and table grapes. Postharvest Biol. Technol. 49 (3), 440–442.

Ryall, A.L., 1963. Proceedings of the Seventeenth National Conference on Handling Perishables, Purdue University, Indiana, USA, March 11–14.

Saleh, R., 2004. Computational Model of a Coupled Heat-Transfer Phenomenon in Bounded Developing Laminar Aerosol Flow. <http://webfea-lb.fea.aub.edu.lb/proceedings/2004/SRC-ME-15.pdf>.

Salunkhe, D.K., Wu, M.T., 1973. Effects of subatmospheric pressure storage on ripening and associated chemical changes of certain deciduous fruits. J. Am. Soc. Hort. Sci. 98, 113.

Salunkhe, D.K., Wu, M.T., 1975. Subatmospheric storage of fruits and vegetables. In: Haard, N. F., Salunkhe, D.K. (Eds.), Postharvest Biology and Handling of Fruits and Vegetables. AVI Publishing Co., Westport, CT, pp. 153–171. (Chapter 13).

Salvador, A., Abad, I., Arnal, L., Martinez-javega, J.M., 2006. Effect of ozone on postharvest quality of persimmon. J. Food Sci. 71 (6), S443–S446.

Sander, R., 1999. Compilation of Henry's Law Constants for Inorganic Species of Potential Importance in Environmental Chemistry. <http://www.mpch-mainz.mpg.de/~Sander/res/henry.html> (version 3, April 8, 1999).

Sangster, J., 1997. Octanol-Water Partition Coefficients: Fundamentals and Physical Chemistry. European Journal of Medicinal Chemistry 1997/32/11/842.

Saraswathy, S., Balasubraamanyan, S., Preethi, T.L., 2010. Postharvest Management of Horticultural Crops. Agrabios, India, pp. 576.

Sarwar, S., 2012. Pak mango export to decline from 1.8 m to 1.2 m tonne this year. Daily Times. Saturday, May 26, 2012.

Schneider, J., Serge Gosset, L.G., Pierre-Lefer, M., 1986. Composition and process for coating paper and cardboard process for preparing the compositions and paper and cardboard so obtained. US Patent 4,600,439.

Schönherr, J., 1976. Water permeability of isolated cuticular membranes: the effect of cuticular waxes on diffusion of water. Planta 131, 159–164.

Schönherr, J., 1982. Resistances of plant surfaces to water loss: transport properties of cutin, suberin and associated lipids. In: Lange, O.L., Nobel, P.S., Osmond, P., Ziegler, H. (Eds.), Encyclopedia of Plant Physiology, vol. 12B. Springer, Berlin/Heidelberg/New York, pp. 153–179.

Schönherr, J., Schmidt, H.W., 1979. Water permeability of plant cuticles. Dependence of permeability coefficients of cuticular transpiration on vapor pressure saturation deficit. Planta 144, 391–400.

Schuerger, A.C., Nicholson, W.L., 2006. Interactive effects of hypobaria, low temperature, and CO_2 atmospheres inhibit the growth of mesophilic Bacillus spp. under simulated Martian conditions. Icarus 185, 143–152.

Schuerger, A.C., Ulrich, R., Berry, B.J., Nicholson, W.L., 2013. Growth of Serratia liquefaciens under 7 mbar, 0°C, and CO_2-enriched anoxic atmospheres. Astrobiology 13 (2), 1–17.

Selwitz, C., Maekawa, S., 1998. Inert gases in the control of museum insect pests. J. Paul Getty Trust.

Serna, L., 2006. Opening the door to CO_2. Nat. Cell Biol. 8, 311–312.

Sharp, A.K., 1985. Temperature uniformity in a low-pressure freight container utilizing glycol-chilled walls. Int. J. Refrig. 8 (1), 37–42, Elsevier.

Sharpe, D., Fan, L., McRae, K., Walker, B., MacKay, R., Doucette, C., 2009. Effects of ozone treatment on Botrytis and Sclerotinia sclerotiorum in relation to horticultural product quality. J. Food Sci. 74, 250–257.

Sharples, R.O., 1974. Hypobaric storage: apples and soft fruit. 1973 Report. East Malling Research Station, Kent, UK.

Sharplin, J., Bhambhani, H.J., 1963. Spiracular structure and water loss from Tribolium confusum Duval and Sitophilus granarius (L.) under reduced pressures. Can. Entom. 95, 352–357.

Shatat, F., Bangerth, F., Neubeller, J., 1978. Effect of three different storage procedures on the production of aroma substances in fruits. Gartenbauwissenschaft 43 (5), 214–222.

Shejbal, J., Tonolo, A., Careri, G., 1973. Conservation of wheat in silos under nitrogen. Ann. Technol. Agric. 22 (4), 773–785.

Shipway, M.R., Shinitzki, M., Bramlage, W.J., 1973. Effect of CO_2 on activity of apple mitochondria. Plant Physiol. 51, 1095–1098.

Siegelman, H.W., Chow, C.T., Biale, J.B., 1958. Respiration of developing rose petals. Plant Physiol. 33, 403–409.

Singh, B.N., Littlefield, N.A., Salunkhe, D.K., 1970. Effect of controlled atmosphere storage on amino acids, organic acids, sugar, and rate of respiration of "Lambert" sweet cherry fruits. J. Am. Soc. Hort. Sci. 95 (4), 458–461.

Sirichandria, C., Wasilewska, A., Viad, F., Valon, C., Leung, J., 2009. The guard cell as a single-cell model towards understanding drought tolerance and abscisic acid action. J. Exp. Bot. 60 (5), 1439–1463.

Sisler, E.C., 1979. Measurement of ethylene binding in plant tissue. Plant Physiol. 64, 538–542.

Slatyer, R.O., 1967. Plant Water-Relationships. Academic Press, London/New York.

Slaughter, D.C., 2009. Methods for management of ripening in mangos. 10 pages. <www.mango. org/mdia/55737/methods_for_managment_or_ripening.pdf>.

Smith, J.R., John, P., 1993a. Activation of 1-aminocyclopropane-1-carboxylate oxidase by bicarbonate/carbon dioxide. Phytochemistry 32, 1381–1386.

Smith, J.R., John, P., 1993b. Maximizing the activity of the ethylene-forming enzyme. In: Pech, J.C., Latché, A., Balagué, C. (Eds.), Cellular and Molecular Aspects of the Plant Hormone Ethylene. Kluwer Academic Publishers, Dordrecht/Boston/London, pp. 33–38.

Smith, W.L., Arnt, L., Bromberg, S.E., Dani, N., El Sayed, M.Y., Foland, L.D., et al., 2008. Microbial control using hypochlorous acid vapor. US Patent 20080003171A1 application.

Soderstrom, E.L., Mackey, B.E., Brandl, D.G., 1986. Interactive effects of low-oxygen atmospheres, relative humidity, and temperature on mortality of two stored-product moths (Lepidoptera: Pyralidae). J. Econ. Entomol. 78 (5), 1303–1306.

Sommer, N.F., Fortlage, R.J., 1996. Ionizing radiation for control of postharvest diseases of fruits and vegetables. Adv. Food Res. 15, 147.

Soni, N.K., 2010. Fundamentals of Botany, vol. -pdf ebook; pdf 3485.qwebooks.com/fundamental-of-botany-volume-2-PDF-6348. Tata McGraw Hill Education Private Limited, pp. 420. Publishing house Private limited, New Delhi.

Spalding, D.H., Reeder, W.F., 1976a. Low-pressure (hypobaric) storage of limes. J. Am. Soc. Hort. Sci. 101, 367–370.

Spalding, D.H., Reeder, W.F., 1976b. Low-pressure (hypobaric) storage of avocados. HortScience 11 (5), 491–492.

Spalding, D.H., Reeder, W.F., 1977. Low-pressure (hypobaric) storage of mangos. J. Am. Soc. Hort. Sci. 102, 367–369.

Spalding, D.H., Davis, P.L., Reeder, W.F., 1978. Quality of sweet corn stored in controlled atmospheres or under low pressures. J. Am. Soc. Hort. Sci. 103 (5), 592–595.

Staby, G.L., Cunningham, M.S., Holstead, C.L., Kelly, J.W., Konjoian, P.S., Eisinberg, B.A., et al., 1984. Storage of rose *Rosa* sp. and carnation *Dianthus caryophyllus* flowers. Ann. Bot. 11, 363–368.

Stenvers, N., Bruinsma, J., 1975. Ripening of tomato fruits at reduced atmospheric and partial oxygen pressures. Nature 253, 532–533.

Stewart, J.K., Aharoni, J., 1983. Vacuum fumigation with ethyl formate. J. Am. Soc. Hort. Sci. 108 (2), 295–298.

Stover, R.H., Frieberg, S.R., 1958. Effect of carbon dioxide on multiplication of *Fusarium* in the soil. Nature 181, 788–789.

Street, H.E., 1963. Plant Metabolism. Pergamon Press, MacMillan Co, New York, NY.

Stumm, W., Morgan, J.J., 1981. Aquatic Chemistry: An Introduction Emphasizing Chemical Equilibria in Natural Waters, second ed. Wiley, New York, NY.

Sun, D.-w., Li, X.-t., 2009. Application of hypobaric storage in chilled meat storage. Food Sci. Tech. 2009-02.

Sutherland, D., Samakovlis, C., Krasnow, M.A., 1996. Branchless encodes a *Drosophila* FGF homolog that controls tracheal cell migration and the pattern of branching. Cell 87 (6), 1091–1101.

Suzuki, T., Masuda, M., Friesen, M.D., Fenet, B., Ohshima, H., 2002. Novel products generated from 2′-deoxyguanosine by hypochlorous acid or a myeloperoxidase-$H_2O_2^-Cl^-$ system; identification of diimino-imidazole and amino-imidazole nucleosides. Oxford J. Life Sci. Nucleic Acids Res. 30 (11), 2555–2564.

Tabak, H.H., Cooke, W.B., 1968. The effects of gaseous environments on the growth and metabolism of fungi. Bot. Rev. 34, 126–252.

Tao, F., Zhang, M., Sun, J., 2003. Preservation of mushroom in storage after vacuum cooling treatment. Agrophysics 19, 293–297.

Theologis, A., Zarembinski, P.W., Oeller, P.W., Liang, X., Abel, S., 1992. Modification of fruit ripening by suppressing gene expression. Plant Physiol. 100, 549–551.

Thimann, K.V., 1955. The Life of Bacteria. MacMillan Co, New York, NY.

Thomas, A.C., Beyers, M., 1979. γ-Irradiation of subtropical fruits. A comparison of the chemical changes occurring during normal ripening of mangos and papayas with changes produced by γ-irradiation. J. Agric. Food Chem. 27 (1), 157–163.

Thomas, H., 1980. Leaf senescence. Annu. Rev. Plant Physiol. 31, 83–111.

Thomas, P., Moy, J.H., 1986. Radiation preservation of foods of plant origin. Part 3. Tropical fruits; baananas, mangoes, and papayas. CRC Crit. Rev. Food Sci. Nutr. 23 (2), 147–205.

Thompson, A.K., 2010. Controlled Atmosphere Storage of Fruits and Vegetables, second ed. CAB International, Wallingford, UK.

Thornton, B.C., Sullivan, W.N., 1964. Effects of a high vacuum on insect mortality. J. Econ. Entomol. 57, 852–854.

Tian, S.-P., Fan, Q., Xu, Y., Wang, Y., Jiang, A.-L., 2001. Evaluation of the use of high CO_2 concentrations and cold storage to control Monilinia fructicola on sweet cherries. Postharvest Biol. Tech. 22 (1), 53–60.

Tiedtke, M., 1987. Parametrization of non-convective condensation processes. European Centre for Medium-Range Weather Forecasts. Lecture Series. pp. 56.

Toler, R.W., Dukes, P.D., Jenkins, S.F., 1966. Growth responses of Fusarium oxysporum f. tracheiphilum in vitro to varying oxygen and carbon dioxide tensions. Phytopathology 56, 183–186.

Tolle, W.E., 1969. Hypobaric storage of mature green tomatoes. USDA Agricultural Research Report 842: pp. 1–9.

Tolle, W.E., 1972. Hypobaric Storage of Fresh Produce. Yearbook of United Fresh Fruit & Vegetable Association, pp. 27, 28, 33, 34, 36, 38, and 43. Shelby Publishing Co., Gainesville, GA.

Torres-Rivera, Z., Hallman, G.J., 2007. Low-dose irradiation phytosanitary treatment against Mediterranean fruit fly (Diptera: Tephitidae). Florida Entomol. 90 (2), 343–346.

Troller, J.A., Bernard, D.T., Scott, V.W., 1984. Measurement of water activity, Compedium of AOAC International, seventeenth ed. AOAC International, Gaithesburg, MD.

Tucker, M.L., Laties, G.G., 1985. The dual role of oxygen in avocado fruit respiration: kinetic analysis in computer modeling of diffusion-affected respiratory oxygen isotherms. Plant Cell Environ. 8, 117–127.

Tzortzakis, N., Singleton, I., Barnes, J., 2007a. Deployment of low-level ozone enrichment for the preservation of chilled fresh produce. Postharvest Biol. Tech. 43, 261–270.

Tzortzakis, N., Borland, A.M., Singleton, I., Barnes, J.D., 2007b. Impact of atmospheric ozone-enrichment on quality related attributes of tomato fruit. Postharvest Biol. Tech. 45, 317–326.

Ullah, H., Ahmad, S., Thompson, A.K., Ahmas, W., Nawaz, A., 2010a. Storage of ripe mango (Mangifera indica L.) Cv. Alphonso in controlled atmosphere with elevated CO_2. Pak. J. Bot. 42 (3), 2077–2084.

Ullah, H., Ahmad, S., Thompson, A.K., Anwar, R., Memon, N.U.N., Nafees, M., 2010b. Effect of "oxygen and carbon dioxide" on the post-harvest management in tree-ripe mango storage. J. Chem. Soc. Pak. 32 (4), 485–491.

Ulrich, R., 1970. Organic acids. In: Hulme, A.C. (Ed.), The Biochemistry of Fruits and Their Products, vol. I. Academic Press, London/New York, pp. 89–118.

Ulrich, R., 1975. Controlled atmosphere storage. Part 2. Physiological and practical considerations. In: Pantastico, Er.B. (Ed.), Postharvest Physiology, Handling and Utilization of Tropical and Subtropical Fruits and Vegetables. AVI Publishing Co., Westport, CT, pp. 186–200.

Uota, M., Garazsi, M., 1967. Quality and display life of carnation blooms after storage in controlled atmospheres. USDA Marketing Research Rept. No. 796, pp. 1–9.

Vahisalu, T., Puzõriova, I., Broshé, M., Valk, E., Lepiku, M., Moldau, H., et al., 2010. Ozone-triggered rapid stomatal response involves the production of reactive oxygen species, and is controlled by SLAC1 and OSTI. Plant J. 62 (3), 442–453.

Valley, G., 1927. The effect of carbon dioxide on bacteria. Quart. Rev. Biol. 3, 209–224.

Valley, G., Rettger, L.F., 1927. The influence of carbon dioxide on bacteria. J. Bacteriol. 14, 101–137.

van den Berg, L., Lentz, C.P., 1978. High humidity storage of vegetables and fruits. HortScience 13, 565–569.

Veierskov, B., Hansen, M., 1992. Effects of O_2 and CO_2 partial pressure on senescence of oat leaves and broccoli miniflorets. New Zealand J. Crop Hort. Sci. 20, 155–158.

Veierskov, B., Kirk, H.G., 1986. Senescence in oat leaf segments under hypobaric conditions. Plant Physiol. 66, 283–287.

Vurma, M., Pandit, R.B., Sastry, S.K., Yousef, A.E., 2009. Inactivation of Escherichia coli O157: H7 and natural microbiota on spinach leaves using gaseous ozone during vacuum cooling and simulated transportation. J. Food Prot. 72 (7), 1538–1546.

Waelti, H., Cavalieri, R.P., Johnson, M., 1992. Humidification and weight loss of apples in CA storage. In: Eighth Annual Washington Tree Fruit Postharvest Conference, Yakima, WA, pp. 45–47.

Wager, H.G., 1973. The effect of subjecting peas to air enriched with CO_2. J. Exp. Bot. 25, 338–351.

Wagstaff, C., Bramke, I., Breeze, E., Thornber, S., Harrison, E., Thomas, B., et al., 2010. A specific group of genes respond to cold dehydration stress in cut Alstroemeria flowers whereas ambient dehydration stress accelerates senescence expression patterns. J. Exp. Bot. 61 (11), 2905–2921.

Walker, D.A., Zelitch, I., 1963. Some effects of metabolic inhibitors, temperature and anaerobic conditions on stomatal movement. Plant Physiol. 38 (4), 390–396.

Wallhäuber, K.H., 1988. Praxis der sterilization-desinfektion-konserving-keim identifizierung-Betriebshygiene. Georg Thieme, Verlag, Stuttgart.

Wang, G.L., Jiang, B.H., Rue, E.A., Semenza, G.L., 1995. Hypoxia-inducible factor 1 is a basic-helix-helix-PAS heterodimer regulated by cellular O_2 tension. Proc. Natl. Acad. Sci. U.S.A. 92, 5510–5514.

Wang, L., Zhang, P., Wang, S.-j., 2001. Advances in research on theory and technology for hypobaric storage of fruit and vegetables. Storage Process (5), 3–6 (in Chinese).

Wang, L., Zhang, P., Wang, S.-j., 2004a. Effect of mini-hypobaric facility on storage weight loss from Dongzao Jujube. Storage Process 2004-04.

Wang, L., Zhang, P., Wang, S.-j., 2004b. Effect of mini-hypobaric facility on storage weight loss from Dongzao Jujube. Storage Process (2), 26–28 (in Chinese).

Wang, L.-x., Zhang, Y.-l., Chen, J.-p., Hao, G.-j., Fu, C.-c., Zhou, R., 2008. Study on preventing Dongzao Jujube fruit Serosa disease with low temperature storage with hypobaric and ozone treatments. Food Sci. 2008-03.

Wang, S.-j., Yan, T.-c., Jia, F.-s., Jiang, Y.-b., 2004c. Study on physiological and biochemical change of *Zizyphus jujuba* in hypobaric storage in West of Liaoning. Storage Process 2004-02.

Wang, W., Zhang, Y.-l., 2008. Control of hypobaric treatment on softening of postharvest apricot fruits. Acta Bot. Boreali Occidentalia Sin. 2008-01.

Wang, W., Meng, Q., Zhang, G., Zhang, P., Liang, Y., Feng, C., et al., 2007. Effect of hypobaric storage on the fruit physiological and biochemical properties of Angelo plum. J. Chin. Inst. Food Sci. Tech. 2007-02.

Wang, X., 1991. Storage of mangos and control of flowers and harvest season. Guangxi Agric. Sci. 3, 110–113.

Wang, Y., 2006. Experimental study on low temperature hypobaric storage of gold pears. Energy Conserv. Tech. 2006-03.

Wang, Z., Dilley, D.R., 2000. Hypobaric storage removes scald-related volatiles during low-temperature induction of superficial scald in apples. Postharvest Biol. Tech. 18, 191–199.

Wankier, D.N., Salunkhe, D.K., Campbell, W.F., 1970. Effects of controlled atmosphere storage on biochemical changes in apricots and peaches. J. Am. Soc. Hort. Sci. 95 (5), 604–609.

Wardlaw, C.W., Leonard, E.R., 1940. Studies in tropical fruits. IX. The respiration of bananas during ripening at tropical temperatures. Ann. Bot. (N.S.) 3, 845–869.

Warner, H.L., Leopold, A.C., 1971. Timing of growth regulator responses in peas. Biochem. Biophys. Res. Comm. 44, 989–994.

Weis-Fogh, T., 1964. Diffusion in insect wing muscle, the most active tissue known. J. Exp. Biol. 41, 229–256.

Welby, E.M., McGregor, B., 1997. Agricultural Export Transportation Handbook. USDA, Agricultural Handbook 700.

Wells, J.M., 1970. Modified atmosphere, chemical and heat treatments to control postharvest decay of California strawberries. Plant Dis. Rep. 54 (5), 431–434.

Wells, J.M., 1974. Growth of *Erwinia carotovora*, *E. atroseptica* and *Pseudomonas fluorescence* in low-oxygen and high-carbon dioxide atmospheres. Phytopathology 64, 1012–1015.

Wells, J.M., Spalding, D.H., 1975. Stimulation of *Geotrichum candidum* by low-oxygen and high carbon dioxide atmospheres. Phytopathology 65, 1229.

Wells, J.M., Uota, M., 1970. Germination and growth of five fungi in low-oxygen and high carbon dioxide atmospheres. Phytopathology 60, 50–53.

Wenxiang, L., Zhang, N., Hab-qing, Y., 2006. Study on hypobaric storage of green asparagus. J. Food Eng. 73, 225–230.

Wheeler, R.M., Mackowiak, C.L., Yorio, N.C., Sager, J.C., 1999. Effects of CO_2 on stomatal conductance: do stomata open at very high CO_2 concentrations?. Ann. Bot. 83 (3), 243–251.

Wheeler, R.M., Wehkamp, C.A., Stasiak, M.S., Dixon, M.A., Rygalov, V.Y., 2011. Plants survive rapid decompression: Implications for bioregenerative life support. Adv. Space Res. 47 (9), 1800–1807.

Wigglesworth, V.B., 1930. Tracheal respiration. Proc. R. Soc. London B106, 229–250.

Wigglesworth, V.B., 1935. The regulation of respiration in the flea, *Xenopsylla cheopsis* Roths. (Pulicidae). Proc. R. Entomol. Soc. London (B) 11, 397–419.

Wigglesworth, V.B., 1954. The Physiology of Insect Metamorphosis. Cambridge University Press, Cambridge.

Wigglesworth, V.B., Gillett, J.D., 1936. The loss of water during ecdysis in *Rhodnius proxlixus* (Hemiptera). Proc. R. Entomol. Soc. London (A) 11, 104–107.

Wilkerson, E., Bucklin, R., Fowler, P., Wheeler, R., Peterson, J., 2004. Design considerations for a greenhouse on Mars: Accounting for plant evapo-transpiration. Paper number 044100, American Society of Agricultural and Biological Engineers Annual Meeting, St. Joseph, MI.

Williams, A.E., Rose, M.R., Bradley, T.J., 1997. CO_2 release patterns in *Drosophila melanogaster;* the effect of selection for desiccation resistance. J. Exp. Biol. 200, 615–624.

Williams, M.W., Patterson, M.E., 1964. Nonvolatile organic acids and core breakdown of Bartlett pears. J. Agric. Food. Chem. 12, 80–83.

Willmer, C.M., Fricker, M., 1996. Stomata, second ed. Chapman & Hall, London.

Willmer, C.M., Mansfield, T.A., 1970. Effects of some metabolic inhibitors and temperature on ion-stimulated stomatal opening in detached epidermis. New Phytol. 69, 983–992.

Willmerr, C.M., 1988. Stomatal sensing of the environment. Biol. J. Linn. Soc. 34, 205–217.

Wills, R.B.H., McGlasson, W.B., Graham, D., Lee, T.H., Hall, E.G., 1989. Postharvest. An Introduction to the Physiology and Handling of Fruit and Vegetables. AVI, Van Nostrand Reinhold, New York, NY.

Wills, R.B.H., Warton, M.A., Mussa, D.M.D.N., Chew, L.P., 2001. Ripening of climacteric fruits initiated at low ethylene levels. Aust. J. Exp. Agric. 41, 89–92.

Wingrove, J.A., O'Farrell, P.H., 1999. Nitric oxide contributes to behavioral, cellular, and developmental responses to low oxygen in *Drosophila.* Cell 98 (1), 105–114.

Wu, A.-x., Zhou, S.-s., Li, W.-x., Jiang, W.-l., Zhang, J., 2011. Effect of different vacuum pressures on the storage of whangkaumbae pears. Food Mach. 2011-01.

Wu, M.T., Salunkhe, D.K., 1972a. Fungi static effects of subatmospheric pressures. Experientia 28, 866–867.

Wu, M.T., Salunkhe, D.K., 1972b. Subatmospheric pressure storage of fruits and vegetables. Utah Sci. 33 (1), 29–31.

Wu, M.T., Jadhav, S.J., Salunkhe, D.K., 1972. Effects of subatmospheric pressure storage on ripening of tomato fruits. J. Food Sci. 37, 952–956.

Xue, M.-l., Zhang, J., Zhang, P., Wang, L., 2003a. Effects of hypobaric storage and treatment with GA-3 on alcoholism of Jujube fruit during cold storage. Chinese Assoc. Refrig. [C] 2003.

Xue, M.-l., Zhang, P., Zhang, J.-s., Wang, L., 2003b. Effects of hypobaric storage on physiological and biochemical changes in postharvest Dong Jujube fruit during cold storage. Sci. Agric. Sin. 36 (2), 196–200.

Yahia, E.M., Rivera, M., Hernendez, O., 1992. Responses of papaya to short-term insecticidal oxygen atmospheres. J. Am. Soc. Hort. Sci. 117, 96–99.

Yang, H., Wu, F., Zhou, C., Wang, Y., 2010. Changes of quality and some enzyme activities of "Dongkui" Chinese bayberry during hypobaric storage. J. Chin. Inst. Food Sci. Tech. 2010-01.

Yang, S.F., Dong, J.G., Fernandez-Maculet, J.C., Olson, D.C., 1993. Apple ACC oxidase: purification and characterization of the enzyme and cloning of its cDNA. In: Pech, J.C., Latché, A., Balagué, C. (Eds.), Cellular and Molecular Aspects of the Plant Hormone Ethylene. Kluwer Academic Publishers, Dordrecht/Boston/London, pp. 59–64.

Yao, Q., Zhang, C., He, J., 2009. The application research development of hypobaric storage in fruits and vegetables. Acad. Period. Farm Prod. Process. 2009-04.

Yi, Z.-M., 2010. Study on postharvest quality deterioration and hypobaric storage of Loquat fruit. Zheijiang University of Technology Masters Thesis.

Yoshida, T., 1975. Nitrogen atmosphere and pest insects. Reports on pest insects in milk powder. J. Food Hyg. Soc. Jpn. 16, 1–11.

Zapotoczny, P., Markowski, M., Majewska, K., 2003. The quality of cucumbers stored under hypobaric conditions. In: Proceedings of the Eighth International CA Conference. Acta Hortic. ISHS 2003, pp. 193–196.

Zelitch, I., 1969. Stomatal control. Annu. Rev. Plant Physiol. 20, 329–350.

Zhang, G.-y., Yang, J.-m., Zhang, P., Liu, Y.-q., 2005b. Effect of low pressure storage on physiological and biochemical changes of Angelo plum fruit during cold storage. Food Sci. 2005-06.

Zhang, M., Yu, H.-q., 2004. Study on hypobaric storage of *Asparagus officinalis*. Wuxi Univ. Light Ind. 23 (6), 38–42 (in Chinese).

Zhang, Y.-Z., Han, J.-Q., Zhang, R.G., 2005a. Study on fresh-keeping physiological activity of Dong Jujube using low temperature combined with hypobaric and ozone treatment. Sci. Agric. Sin. 38 (10), 2102–2110.

Zheng, X., Xiong, Y., 2009a. The practice of free from freezing preservation of fresh produce and food safety on hypobaric storage technology. Chinese Assoc. Refrig. [C] 2009.

Zheng, X., Xiong, Y., 2009b. Primary experiment investigating fresh free from freezing preservation by hypobaric cold storage using continuous ventilation. Chinese Assoc. Refrig. [C] 2009.

Zheng, X., Zheng, Q., 2008. Hypobaric storage technology cannot be substituted by other technology in non-frozen fresh preservation, storage and transport of fresh produce. Chinese Assoc. Refrig. [C] 2008.

Zheng, X., Zheng, Q., 2009a. Progress on preservation fresh technology of research on hypobaric storage. Chinese Assoc. Refrig. [C] 2009.

Zheng, X., Zheng, Q., 2009b. Effect of hypobaric storage on the preservation of cucumbers. Storage Process 2009-04-11.

Zheng, X., Zheng, X., 2009c. Multifunctional decompressing storage device. Chinese Patent Application No. 2009100510253, Authorized Announcement No. 101554941B.

Zheng, X., Jiang, L.-J., Xiong, E.-Y., 2011. Research and application of free from freezing preservation fresh technology of hypobaric storage. Chinese Society of Agricultural Engineering [C] 2011.

Zhou, G., Liu, Z., 2004. Simulating study on heat and mass transfer of food in hypobaric and refrigerated preservation. Trans. Chin. Soc. Agric. Mach. 2004-01-027.

Zhou, H.j., Qiao, Y.-j., Zhang, S.-l., Wang, H.-h., Chen, Z.-l., Zheng, X.-z., 2010. Effects of low temperature and hypobaric storage on softening and membrane injury physiology of Datuanmilu peach. Jiangsu J. Agric. Sci. 2010-04.

Zhou, Y., Gao, H., Chen, H., Mao, J., Chen, W., Song, L., 2008. Effects of hypobaric storage on active oxygen metabolism of persimmon (*Diospyros kaki* L.) fruit. ISHS Acta Hortic. 804, 523–526.

Zhu, H., Blackborow, P., 2011. Operation of Laser-Driven Light Sources Below 300 nm: Ozone Mitigation. Energetiq Technology, Inc., Application note, pp. 1–3.

Zhu, J.-y., Chai, X.-s., 2005. Some recent developments in headspace gas chromatography. Curr. Anal. Chem. 1, 79–83.

Zou, S.-s., Wu, A.-x., Li, W.-x., Wang, S.-q., Jiang, W.-l., 2011. Effect of microvacuum storage on the fresh-keeping of Laiyang pear. Food Sci. 2011-12.

Zuckerman, H., Harren, F.J.M., Reuss, J., Parker, D.H., 1997. Dynamics of acetaldehyde production during anoxia and post-anoxia in red Bell pepper studied by photo-acoustic techniques. Plant Physiol. 113, 925–932.

Note: Page numbers followed by "f" and "t" refer to figures and tables, respectively.

Academic Press is an imprint of Elsevier
32 Jamestown Road, London NW1 7BY, UK
The Boulevard, Langford Lane, Kidlington, Oxford, OX5 1GB, UK
Radarweg 29, PO Box 211, 1000 AE Amsterdam, The Netherlands
225 Wyman Street, Waltham, MA 02451, USA
525 B Street, Suite 1900, San Diego, CA 92101-4495, USA

First published 2014

British Library Cataloguing-in-Publication Data
A catalogue record for this book is available from the British Library

Library of Congress Cataloging-in-Publication Data
A catalog record for this book is available from the Library of Congress

ISBN: 978-0-12-419962-0

For information on all Academic Press publications
visit our website at **store.elsevier.com**

Working together
to grow libraries in
developing countries

www.elsevier.com • www.bookaid.org

Printed in the United States
By Bookmasters